CATTLE WEALTH

A Journey from Gaumata to Menace to Violence

CATTLE WEALTH

A Journey from Gaumata to Menace to Violence

By

Dr. K. P. Agrawal
Independent Consultant
and former National Coordinator
National Agricultural Innovation Project
Indian Council of Agricultural Research
New Delhi-110001

Worldwide Published by
Pendown Press

PENDOWN PRESS

An ISO 9001 & ISO 14001 Certified Co.
Regd. Office 2525/193, 1st Floor, Onkar Nagar-A,
Tri Nagar, Delhi-110035
Ph.: 09350849407, 09312235086
E-mail: info@pendownpress.com
Branch Office 1A/2A, 20, Hari Sadan, Ansari Road,
Daryaganj, New Delhi-110002
Ph.: 011-45794768
Website: PendownPress.com

First Edition: 2018
Price: ₹299/-
ISBN: 978-93-88149-41-9

Layout and Cover Designed by Pendown Graphics Team
Printed and Bound in India by Thomson Press India Ltd.

Contents

Foreword....vii
Preface....ix
Acknowledgements....xi
About the Author.... xiii
1. Introduction....1
2. Cow and Religion....11
3. Cow - a Moving Wealth....25
4. Cow Menace....41
5. Vigilantism and Violence....55
6. Cow Shelters are Becoming Cow Killers....73
7. Conserving Indigenous Cattle Breeds....89
8. Improving Feed and Fodder Resources....111
9. Scientific Approach of Cattle Keeping to Improve productivity and Profitability....129
10. Way forward To Revive Cattle Wealth....139

Annexure-1

Acronyms....151

आईसीएआर–भारतीय पशु चिकित्सा अनुसंधन संस्थान
(सम–विश्वविद्यालय)
इज्जत नगर - 243 122 (उ.प्र.) भारत

भाकृअनुप
ICAR

ICAR-INDIAN VETERINARY RESERCH INSTITUTE
[Deemed University]
Izatnagar – 243 122 [U. P.] INDIA

IVRI
ivri

डॉ आर.के. सिंह
Dr R.K. Singh
निदेशक एवं कुलपति
Director & Vice-Chancellor

FOREWORD

Since time immemorial, India has been an agrarian country and cow has been its backbone. It plays an important role in socio-economic life of people in general and in villages in particular. The livestock sector contributes about 4.1% to the total GDP of the country. Cow has a long history and has been the favourite to Lord Krishna.

Many important issues related to cow welfare, which the cow lovers across the country are facing for long have been addressed in this book. Some of the genuine issues need immediate attention of the concerned authorities. I congratulate Dr K P Agarwal for his hard work to make this important manuscript available to the country.

27/8/18

(R K Singh)

Phone: 0091-581-2300096(O): 2302231 (R): Fax: 0091-581-2303284; E.mail: director.ivri@icar.gov.in, directorivri@gmail.com

On the setoice of nation since 1889

स्वच्छ भारत
एक कदम स्वच्छता की ओर

Preface

The Book, packed in 10 chapters, provides in-depth and detailed information on different aspects of Cow Wealth. The Issues such as: Cow And Religion, Cow - A Moving Wealth, Cow Menace, Vigilantism And Violence, Cow Shelters Are Becoming Cow Killers, Conserving Indigenous Cattle Breeds, Improving Feed And Fodder Resources, Scientific Approach to improve productivity and Way forward To Revive Cattle Wealth, which the cattle lovers across the country are facing for long have been covered in the book. Cow journey from 'Gaumata to Menace to Violence' is a big setback to the country. Rural economy (particularly dairy farmers) is badly affected. Violence in the name of cow protection has hurt sentiments of dairy farmers and is a shame to the country. The situation is becoming bad to worse to worst over the years. Time is to wake up now.

The book will be of immense value to policymakers, programme planners, public and private sectors, NGOs, social workers, environmental workers, educationists, developmental departments and dairymen who never dream to see cow on street feeding on garbage as "A Gaumata of Ancient India".

Acknowledgements

The sources of information in the manuscript are important articles from National Dailies like Hindustan Times, Times of India, Economic Times, Navbharat Times (Hindi); Magazines like Outlook, Down to Earth, the Speaking Tree, Dwarka Forum, the reports of various committees, commissions, developmental programmes, media, internet, personal communication and on the spot observations by the author. Valued inputs from different sources are gratefully acknowledged. My grateful thanks are also to different organisations, institutions, scientists, social and extension experts working on different aspects of livestock developments with particular reference to cow in the country. Without their support, it would not have been possible. It is also a pleasure to acknowledge the help rendered by Sanjay Singh for typing the manuscript. Last but not the least, I express my heartfelt appreciation to my wife Shakuntala, my daughter-in-law Parul, my son Jayesh, my daughters Rachna and Rani, son-in-laws Anuj and Tarun and grandsons Atharv, Naman, Kushagra and Parth and grand daughters Saaksha and Adya for their whole hearted support and encouragement in writing the book.

About the Author

Dr. K. P. Agrawal obtained his Bachelor of Veterinary Science and Animal Husbandry (1968) and Master of Veterinary Science (1972) digress from the Veterinary College, Mathura and Doctorate (1980) from Indian Veterinary Research Institute (I.V.R.I.), Izatnagar. He had a humble beginning of his research/teaching career as a demonstrator in the Physiology Department at the Veterinary College, Mathura (December 1968 to September 1970). Since then, he worked in different capacities: as research assistant; senior research assistant; scientist; senior scientist at IVRI, Izatnagar; associate research scientist/associate professor/research scientist at the Veterinary College, Gujarat Agricultural University, Gujarat; and scientist-S3/principal scientist/head of the division at the Central Institute for Research on Goats, Makhdoom. He was selected as the director of the Department of Biotechnology, the Ministry of Science and Technology, New Delhi, in 1990. He has 38 years of service experience, of which 30 years of wide experience in research, teaching, extension, research management, policy planning and institutional building. His significant research contribution is in the area of embryo transfer technology. He assumed the responsibility of the National Coordinator of the World Bank-aided National Agricultural Technology Project. (NATP) on 1st April, 1999. As a follow-up of the NATP, he continued as the National Coordinator of another World Bank-aided project, the National Agricultural Innovation Project (NAIP) before he retired in July 2007. He has 350 publications to his credit, of which 30 are books, chapters in books, monographs and bulletins. Dr. Agrawal has been the recipient of several awards/honours during his academic and research career.

After retirement, Dr. K.P. Agrawal worked as a Senior Consultant in World Bank-Supported Projects, such as the "Water Sector Restructuring Project" in Madhya Pradesh, "Rajasthan Agricultural Competitive Project" in Rajasthan and "National Agricultural Innovation Project" in Indian Council of Agricultural Research (ICAR). Dr. Agrawal also authored two books (1. Water "The Elixir of Life" is endangered and 2. Delhi- A Role Model of Urban India) after retirement. Currently he is an independent consultant in several research and developmental programmes in agriculture and allied sectors in the country.

1

Introduction

In India, animal husbandry and agriculture are intrinsically linked, each one being dependent on the other and both crucial for rural livelihood and food security. Livestock production provides livelihood security for most of the rural population (particularly to farmers). It contributes to the nutrition of the households, offering employment opportunities, supplementing incomes and supporting agriculture in the forms of inputs (fertiliser and draught power for agriculture operations).

India's 20 million people are working in animal farming sector. The sector plays a multi-faceted role in the socio-economic development of the country. It supports more than two-thirds of Indian rural population and contributes an estimated 27% to the national agricultural GDP. India is the largest producer of milk, accounting for 17% of the total milk in the world (Source: Agriculture Today - August 2016). However, the milk productivity (production per animal) in the country is far less as compared to those in developed dairy nations.

Cattle—colloquially cows—are the most common type of large domesticated ungulates. They are a prominent modern members of the subfamily Bovinae. They are the most widespread species of the genus Bos, and are most commonly classified collectively as Bos Taurus.

The cattle being more than 70% of large ruminant population in the country contributes significantly to reduce poverty and ensure food and nutritional security. Cattle contribute to the livelihoods of more than two-thirds of the rural poor. Cattle can reduce poverty through three approaches namely; 1) securing the assets and other resources, 2) improving the productivity of agricultural systems and 3) encouraging participation of the poor in livestock-related markets.

The average daily milk yield for crossbred cattle in India is 7.1 kg per day while it is 25.6 kg in UK, 32.8 kg in USA and 38.6 kg in Israel. The reason behind the low yield in India is because of both intrinsic and extrinsic factors. The intrinsic factor is low genetic potential while extrinsic is related with number of reasons like poor nutrition and feed management, inferior farm management practices and inefficient implementation of breed improvement programmes. At present, India is largely using Jersey, a native of Netherlands and British origin Holstein for cross-breeding purposes. Amazing facts about dairy farming in India are mentioned below.

Amazing Facts about Dairy Farming in India

(1) India is the world's biggest producer and consumer of dairy products. Milk production- 140bn litres a year (UK farmers delivered 14.4bn litres in 2014-15) and the country is self-sufficient in dairy, with demand growing 4.5% year on year.

(2) Of that 140bn litres, about 50bn litres are kept by the farmers for their own consumption.

(3) India has 300 million bovines. One-third are buffaloes, which provide 55% of the milk.

(4) In a country of 1.3 billion people, the average herd size is between one and two cows.

(5) The average net income for a dairy farmer is US$387 (£247) a cow each year. That is not much when you've only got one or two animals.

(6) Dairy farming is mainly a women's game. They look after the livestock while men typically find a wage beyond the home.

(7) Indian farmers receive 80-86% of every rupee spent on milk by consumers. This is due to supply chain efficiency and very few big retailers. British farmers' share of the on-shelf price is about 36%.

Cattle rearing has been a traditional livelihood in India and is closely linked to agricultural economy. India with 199 million cattle has 14.5% of the world cattle population. Of this, 83% i.e. 166 million are indigenous. Most of the indigenous cattle (about 80%) are non-descript and only 20% belong to breeds recognised by National Bureau of Genetic Resources. The cattle genetic resource of India is represented by 30 well recognised indigenous breeds. Indigenous cattle, in India, are robust and resilient and are particularly suited to the climate and environment of their respective breeding tracts. They are endowed with qualities of heat tolerance, resistance to diseases and the ability to thrive under extreme climatic stress and less than optimal nutrition. Features of indigenous cattle are mentioned below.

Features of Indigenous Cattle

(1) Cattle are the first animals domesticated by man for agricultural purpose.

(2) Brahman breed in USA introduced in 1849 has been developed from Indian cattle breeds.

(3) Indian cattle breeds like Gir and Ongole (Nellore) landed on Brazilian shores in 1906 are thought to be the world's oldest domestic cattle breeds and can be seen all over Brazil whereas, in their native country, India, these breeds are rarely found.

(4) Cow was a second outcome (Ratna) of "Samudhra Manthan".

(5) A great medicine called "Panchagavya" (cow milk,cow ghee, cow curd, cow urine and cow dung) was made from indigenous cow by the sage, "Dhanwanthri".

(6) The Indian breeds excel in draught capacity, exhibit distinct superiority in utilizing poor quality feed (grass, crop residues and crop by-products) and are adapted to adverse climate.

(7) Cattle produce fuel, fertiliser, nutrition and generates more power than all of India's generating plants. Eight hundred million tones of manure produced every year is the biggest energy source. To replace dung with coal would cost India $1.5 billion per year.

(8) Cow dung if packed on the outside walls of a house, keeps the house warm in winter and produces a cooling effect in summer. It also keeps flies away, and its smoke acts as a repellent for mosquitoes.

(9) India is first in milk production. Average milk yield varies from 1800 to 3500 kg per lactation.

(10) Before mechanisation in agriculture, India depends on bullock power. Bullocks used to do two-thirds of the work on an average farm. Some fifteen million bullock carts were used to move approximately fifteen billion tones of goods across the nation. Converting Indian agriculture from bullocks to machinery has costed to the

country an estimated thirty billion dollars plus maintenance and replacement costs.

India has rich and diverse genetic resources, with some of the best breeds of dairy, draught and dual purpose cattle. Over 30 indigenous cattle breeds are essentially the result of long-term natural selection and evolution over the centuries and are fully adapted to different agro-eco-regions of the country, withstand tropical diseases and perform reasonably well even with low inputs in terms of feed and fodder. Indigenous cattle breeds are declining at a faster rate for example, in Andhra Pradesh, Ongole pure-bred population is decreasing and the Punganur breed has almost vanished. Similarly, Sahiwal population has dwindled in Haryana and Amritmahal in Karnataka. In Kerala, the Vechoor cow is almost on the verge of extinction. The possible reasons for the decline in the population are inadequate inputs in terms of feed/fodder, shelter and management, poor reproductive management, dilution of breeds due to inter-mixing, indiscriminate infusion of exotic germplasm for cross-breeding for enhancing milk production, absence of specific strategies and programmes for conservation of indigenous breeds and loss of breeds due to geographical re-organisation. Breeding tracts and organised farms of some of the established breeds such as Sahiwal, Red Sindhi and Tharparkar have been lost to Pakistan. On account of the prospects of quick economic gains, cross-breeding has encroached into even the native tracts of superior indigenous breeds, which has led to dilution of these breeds, some of which are threatened and heading toward extinction if no action is taken to arrest this trend. Even in the cross-bred strains, genetic stabilisation is yet to take place. The male calves of crossbred cattle are generally considered by the farmer inferior as they are not useful as draught animals. These calves are neglected and starved to death, or more often than not, sold for slaughter for meat and hides, illegally in most states. These calves cannot be slaughtered due to social and ethical reasons in few states of India.

India is rich in genetic diversity and possesses some of the best breeds of cattle in the world. Our indigenous cattle breeds, such as Gir, Ongole (Nellore) are thought to be the world's oldest cattle. Zebu cattle bullocks were introduced in the United States as early as 1849. Brahman breeds in USA have been developed from Indian cattle breeds. First Indian cattle breeds (like Gir and Ongole) landed on Brazilian shores in 1906 are the finest Zebu specimens. The country's population of 218.8 million accounts for 17% of the total world population of cattle. The best indigenous germplasm of milch, draught and dual purpose animals accounts for 22-25% of the Indian cattle population, while 8-10% of the cattle population is cross-bred. Remaining (65-70%) are non-descript. Indigenous cattle breeds excelling in draught capacity, exhibiting a distinct superiority in utilizing poor quality feed and resistant for several tropical diseases and are fully adapted to Indian conditions produce milk varying from 1800 to 3500 kg per lactation. Indigenous cattle breeds of India, their breeding tracts and important features are mentioned below.

Cattle Breeds of India

Breed	Breeding Tracts	Important Features
Gir	Saurashtra region of Gujarat	2000 to 6000 kg milk, bullocks are powerful draught animals
Sahiwal	Punjab, Uttar Pradesh and Haryana	2000 to 4000 kg milk
Red Sindhi	Sind area of Pakistan, Southern part of Gujrat, Rajasthan	2000 to 4000 kg milk
Rathi	Bikaner and	Resistance to adverse

	Ganganagar (Rajasthan), Sirsa (Haryana) and Ferozepore (Punjab)	climatic conditions of the desert, 1800 kg to 3500 kg milk
Tharparkar	Kutch, Jodhpur and Jaisalmer	1800 to 3500 kg milk
Haryana	Haryana, U.P, Bihar and parts of Rajasthan	Dual purpose breed, bullocks are good for draught, 1000 to 2000 kg milk
Kankrej	Bani tract of Bhuj District, North Gujarat and part of Rajasthan	Bullocks are strong and hard-working, 1500 to 4000 kg milk
Deoni	Marathwada region of Maharashtra and adjoining area of Karnataka and AP	Dual-purpose breed
Ongole	Guntur, Ongole (AP)	Bullocks are useful for ploughing and carting, cows are good milker
Mewati, also known as Kosi	West Alwar and Bharatpur districts of Rajasthan	Sturdy, powerful, docile and are good for heavy ploughing, carting and drawing water from deep wells, cows are good milkers
Red Kandhari	Marathwada region of Maharashtra	Dual purpose breed, bullocks are hardy
Nimari	Khandwa (MP)	Dual purpose breed
Malvi	Madhya Pradesh and Rajasthan	Dual purpose breed
Gaolao	Vidarbha Region of Maharashtra	Bullocks are useful for ploughing, cows are average milkers
Dangi	Western Ghat	Bullocks are good for

	Region of Maharashtra	draught purpose in heavy rainfall area and hilly tract
Khillar	Southern part of Maharashtra	Bullocks are hardy and well-known for being fast in work
Amritmahal	Karnataka	Bullocks are known for draught power and endurance, 1000 to 1200 kg milk
Hallikar	Hassan, Tunkur (Karnataka)	Draught breed
Nagori	Nagore (Rajasthan)	Excellent draught breed
Kenkatha	Ken river of Banda District of U.P. and M.P.	Bullocks are small but fairly sturdy and good for cultivation in rocky areas
Siri	Hill tracts around Darjeeling and Sikkim. Bhutan is home tract of this breed	Bullocks are used for draught purpose in rugged conditions of mountains
Bachaur	Sitamarhi (Bihar)	Bullocks are powerful and used for draught purpose
Kherigarh	Lakhimpur Kheri District of U.P.	Bullocks are good for light draught and transport work, animals are active and can thrive on grazing only
Krishna Valley	Southern border of Maharashtra and Andhra Pradesh	Bullocks are powerful, good for heavy ploughing and slower draught work, cows are fair milkers
Kangayam	Coimbatore (Tamil Nadu)	Bullocks are strong draught animals
Bargur	Coimbatore (Tamil Nadu)	Bullocks are good draught animals
Umblachery	Thanjavur District of Tamil Nadu	Draught breed similar to Kangayam

Bargur	Erode district of Tamil Nadu	Bullocks are suitable for agricultural operations in hilly terrains
Ponwar	Pilibhit and North West part of Lakhimpur Kheri District of UP	Draught purpose small animals
Vechoor	Kerala	Bullocks are mainly used for draught purpose, cows are poor milkers

Cow has a long history. In Hinduism, its status was raised to the level of Goddess. They were the favorite animals of Lord Krishna. Cow, a symbol of wealth, strength, and abundance during Lord Krishna's time, is now destitute and a cause of violence. The Issues such as: Cow And Religion, Cow - A Moving Wealth, Cow Menace, Vigilantism And Violence, Cow Shelters Are Becoming Cow Killers, Conserving Indigenous Cattle Breeds, Improving Feed And Fodder Resources, Scientific Approach to improve productivity and Way forward To Revive Cattle Wealth, which the cattle lovers across the country are facing for long have been covered in the book.

2

Cow and Religion

The earliest Vedas do not prohibit the slaughter of cattle. In fact, they ordain it as a part of sacrificial rites. Cattle were not only the major sacrificial victims in the Vedic times, but cow milk products like ghee were also regularly used for oblation -- a practice that continues to this day. Moreover, the Brahmans that performed the cow sacrifices also readily ate the consecrated beef. It is true that the Rigveda (1500-1200 BCE) alludes to the mystical relationship between the cow and the universe, while the Atharvaveda (1200-1000 BCE) calls the cow the "all producing and all containing universe" but they are silent concerning matters of cow slaughter.

In the Dharmaśāstra, Vasistha states that a host may offer hospitality to a Brahman priest by cooking a full-grown ox, while Gautama notes that if an ascetic invited to eat at a sacrifice rejects meat then he shall go to hell for as many years as the slaughtered beast has hairs. It is not until the very end of the Vedic period that we find even the slightest allusions to cow slaughter. Therefore, it is safe to assume that at the close of the Vedic period, the cow was still being slaughtered and eaten by Hindus, even though it served as a powerful symbol. The Bhagavata Purana also allows the consumption of meat in emergency situations like famines. In fact, it prohibits alcohol more than it does meat.

An ancient text, "Charaka Samhita" also stated the advantages of eating cow meat apart from various other kinds of animal meat. The text stated that beef was good for those suffering from loss of flesh due to disorders such as rhinitis, dry cough, fever and fatigue. In ancient India, beef-eating was perfectly normal. Caste was based on profession and if you changed your profession, you changed your caste. Consolidation of castes came only later. There is no injunction in books that so and so can not eat beef. The ban on beef is clearly un-scientific and is not supported by the Vedas, Shastras or the Upanishads (Source: Outllook-07.12.2015).

However, cows are considered sacred by Hindus in India. They were the favorite animals of Lord Krishna, and they serve as a symbol of wealth, strength, and abundance. Millions of Hindus revere and worship cows. Hinduism is a religion that raises the status of cow to the level of Goddess. The cow is a symbol of the divine bounty of earth. Lord Krishna, one of the most well known of the Hindu deities is often depicted playing his flute amongst cows and dancing Gopis (milkmaids). Throughout the Vedic scriptures there are verses which emphasize that the cow must be protected and cared for. It is considered a sin to kill a cow and eat its meat. Even today in India, there are many States in which the slaughter of cows is illegal.

History of Cattle slaughter in India

Cattle slaughter is a controversial topic in India because of the cattle's traditional status as an endeared and respected living being to many in Hinduism, Sikhism, Jainism and Buddhism, in contrast to cattle being considered as a religiously acceptable source of meat by many in Islam, Christianity as well as some in Hinduism and other Indian religions. More specifically, the cow's slaughter has been shunned because of a number of reasons such as being associated with God Krishna in Hinduism, cattle being respected as an integral part of rural livelihoods and an essential economic necessity. Historically, cattle slaughter has also been opposed by various Indian religions because of the ethical principle of Ahimsa (non-violence) and the belief in the unity of all life.

The protection of the cow already had a political significance before the British period. The Mughal emperor Akbar banned the killing of cow. After the collapse of the Mughal Empire, cow slaughter was a capital offense in many Hindu and Sikh ruled regions of the subcontinent. The East India Company continued the ban on cow slaughter in many domains. In the 1857 revolt, the Muslim emperor Bahadur Shah II threatened to blow any Muslim caught sacrificing a cattle during Bakr-Id. The independence leader of India, Mahatma Gandhi, championed cow protection. The history of cow slaughter seventieth century onwards is given below.

Before 1800s: Cow slaughter has been punishable by death in many instances in Indian history. Under the Scindia of Gwalior state and the Sikh Empire, people were executed for killing cows. The "Holi Riot" of 1714 in Gujarat was in part cow-related. A Hindu had attempted to start the spring festivities of Holi by burning a public Holika bonfire, a celebration that his Muslim neighbors objected to. The Muslims retaliated by slaughtering a cow in front of Hindu's house. The Hindus gathered, attacked the Muslims, seised the Muslim butcher's son and killed him. Markets and homes were burnt down. Many Hindus and Muslims died during the Holi riot. The cycle of violence continued for a few days devastating the neighborhoods in Ahmedabad.

1800s: In the 1870s, cow protection movements spread in the Punjab, the North-West provinces, Awadh and Rohilkhand. The first Gaurakshini Sabha (cow protection society) was established in the Punjab in 1882. The movement spread rapidly all over North India and to Bengal, Bombay, Madras and other central provinces. The organisation rescued wandering cows and reclaimed them to groom them in places called gaushalas (cow refuges). Charitable networks developed all through North India to collect rice from individuals, pool the contributions, and re-sell them to fund the gaushalas. Arya Samaj and its founder Dayananda Saraswati were one of the early supporters of the cow protection movement.

The first cow protection societies on the Indian subcontinent were started by Kukas of Sikhism, a reformist group seeking to purify Sikhism. The Sikh Kukas or Namdharis were agitating for cow protection and viewed cow protection as a "sign of the moral quality of the state. Widespread cow protection riots occurred repeatedly in British India in the 1880s and 1890s. These were observed in regions of Punjab, United Provinces, Bihar, Bengal, Bombay Presidency and in parts of South Myanmar (Rangoon). The anti-Cow Killing riots of 1893 in Punjab caused the death of at least 100 people.The 1893 cow killing riots started during the Muslim festival of Bakr-Id, the riot repeated in 1894, and they were the largest riots in British India after the 1857 revolt.

1900-1947: Cattle protection-related violence continued in the first half of the 20th century. Examples of serious cow protection agitation and riots include the 1909 Calcutta riot after Muslims sacrificed a cow in public. In 1916 and 1917, over the Muslim festival of Bakri-Id, two riots broke out in Patna with widespread rioting, looting and murders in major cities of Bihar. In the 1920s, over 100 riots, 450 deaths and 5,000 injuries were recorded in Bengal which was divided in 1947 into East Pakistan and West Bengal. In 1946, rumors spread in Bengal that Hindus had secretly conspired to stop cow sacrifice on Bakri Id by bringing in Sikhs and arms into their homes. Violence continued for about a week with "frequent instances of stray killings" and looting.

1947-2014: After the 1947 Partition of the Indian subcontinent into Pakistan and India, frequent riots and fatal violence broke in newly created India over cow slaughter. Between 1948 and 1951, cow slaughter led to a spate of riots in Azamgarh, Akola, Pilibhit, Katni, Nagpur, Aligarh, Dhubri, Delhi and Calcutta. Riots triggered by slaughter of cows continued in rural and urban locations of India in the 1950s and 1960s. According to Ian Copland and other scholars, it was the practical stop of cow sacrifice ritual as Islamic festivals after 1947 that largely led to a reduction in riots from the peak observed just before India's independence. However, they add, the riots re-emerged in the 1960s, when a new generation of Muslims born after the independence reached adolescence, who were less aware of the trauma of religious violence in India of the 1940s, began to assert their rights.

In 1966, 100 members of Indian parliament signed a petition for a nationwide ban on cow slaughter. Hindu sadhus (monks) gathered in Delhi to protest against cow slaughter, launched go-raksha (cow protection) agitation and demanded a ban. During a huge procession that was walking towards the parliament to press their demand, before they could reach the parliament, some people began a disturbance and rioting started. These riots killed eight people. However, the policy of no national ban on cow slaughter was not accepted by the government.

Article 48 of the Constitution of India mandates the state to prohibit the slaughter of cows and calves and other milch and draught cattle. On 26 October 2005, the Supreme Court of India, in a landmark judgement upheld the constitutional validity of anti-cow slaughter laws enacted by different state governments in India. 24 out of 29 states in India currently have various regulations prohibiting either the slaughter or sale of cows. Kerala, West Bengal, Arunachal Pradesh, Mizoram, Meghalaya, Nagaland, Tripura and Sikkim are the states where there are no restrictions on cow slaughter. As per meat export policy in India, the export of beef (meat of cow, oxen and calf) is prohibited. Bone in meat, carcass, half carcass of buffalo is also prohibited and is not permitted to be exported. Only the boneless meat of buffalo, meat of goat and sheep and birds is permitted for export.

Post-2014: There has been a rise in cow protection related violence since in 2014. The violence has included notable killings, such as the lynchings at Dadri, Jharkhand and Alwar. According to the Human Rights Watch, the post-2014 violence has included instances of assault, harassment, extortion, and it has targeted Muslims and lower-caste Hindus. According to a Reuters report, citing IndiaSpend analysis, a total of "28 Indians - 24 of them Muslims - have been killed and 124 injured", between 2010 and June 2017 in cow-related violence.

Cow slaughter ban initiatives

In 1955, a bill was drafted for India's parliament for a nationwide ban on cow slaughter, stating that a "large majority of the party" was in favour. India's first Prime Minister Jawaharlal Nehru opposed this national ban on cow slaughter, and threatened to resign if the elected representatives passed the bill in India's parliament. The bill failed by a vote of 95 to 12. Nehru declared that it was individual states to decide their laws on cow slaughter. State legislatures such as those of Uttar Pradesh, Bihar, Rajasthan and Madhya Pradesh enacted their own laws in the 1950s. In 1958, Muslims of Bihar petitioned the Supreme Court of India that the ban on cow slaughter violated their religious right. The Court unanimously rejected their claim. In 1966, cow protection groups held large demonstrations in Delhi. In the violence, eight people were killed, including one policeman. In March 2015, Maharashtra passed stricter legislation expanded its ban to the slaughter of bulls and bullocks. On 26 May 2017, the Ministry of Environment of Central

Government imposed a ban on the sale and purchase of cattle for slaughter at animal markets across India, under Prevention of Cruelty to Animals statutes, although Supreme Court of India suspended the ban on sale of cattle in its judgement in July 2017, giving relief to the multi-billion dollar beef and leather industries.

Cow Slaughter and Fundamental Rights

Nearly 70 years ago, the Constituent Assembly saw a lively debate on November 24, 1948, on the form in which ban on cow slaughter was to be included in the Constitution. The opinions- whether it should be part of fundamental rights or be included in directive principles of state policy-differ. The majority agreed to make it a part of directive principles. After the constitution was adopted, several states- Bihar, UP, Assam, Bombay, West Bengal, Central Provinces & Berar and Hyderabad - enacted laws in the 1950s to ban cow slaughter and regulate cattle slaughter. No exception was made in these laws to permit slaughter of cattle even for bonafide religious purposes (Source: TOI-01.05.2017).

The ban was challenged in the Supreme Court by butchers, tanners, gut merchants, curers and cattle dealers on the ground that it violated their fundamental right to do business (Article 19(1)(g) and religion (Article 25). The cases reached in the SC. A five judge bench of the SC in Mohd Hanif Quareshi (1958 AIR 731) case partly upheld the validity of the laws, as far as they banned slaughter of cows and calves. However, it said these laws were "void in so far as they totally prohibit the slaughter of breeding bulls and working bullocks without prescribing any test or requirement as to their age or usefulness. Nearly half a century later in 2005, a seven-judge bench of the SC re-evaluated the 1958 judgment and held that the legislature would be well within its right to impose a complete ban on slaughter of cow and its progeny. They can slaughter animals other than cow progeny and carry on their business activity. It is not necessary that the animal must be slaughtered to avail these things (hides, skins, etc). The animal, whose slaughter has been prohibited, would die a natural death even otherwise and in that case their hides, skins and other parts of body would be available for trade and industrial activity.

Ban on Sale of Cattle: May 23, 2017 notification banning sale of cattle for slaughter in livestock markets sparked protests in many parts of the country. The animals which fall under the purview of the notification are cows, bulls, bullocks, buffaloes, steers, heifers, calves and camels. The inclusion of buffaloes was strongly protested by meat exporters and even state governments like Kerala and West Bengal. There were many Hindus as well as people from other religious groups who eat beef. There seems to be no rationale or legality in a section of people imposing their dietary habits on the others in a democratic society. A nationwide ban on cow slaughter would adversely affect the country's economy. It will cripple meat exports worth crores of rupees, apart from adversely impacting the leather and several allied industries. After several cold-blood murders of cattle traders, a controversial ban on cattle slaughter and protests in some states against the beef ban, the Centre has withdrawn the controversial rules that banned the sale of cattle including buffaloes for slaughter in animal markets, as per government notification on 2.12.2017.

Impact of Ban on Sale of Cattle by Meat Traders, Farmers and Industries: Blanket ban on sale of cattle is not justified. Rather focus should be to preventing animal cruelty. Curbs on interstate transport of cattle- already a risk business due to cow vigilantism will hurt the interest of farmers procuring animals from other states. Cows from Haryana and Punjab are in demand in other parts of the country for their superior milk production and disruption of such trade is problematic for all states.

The environment ministry, which had notified the rules on 23 May, 2017, received 13 representations initially but the number has swelled since. The decision has been protested by meat exporters who have said that the move would affect meat business and leather trade. The rules prohibit animal markets within 25km of a state border and within 50km of an international border.

The law will also have a harmful impact on farmers and industries which use livestock products as input. In addition, it undermines federalism as it encroaches on the domain of states. The Trade that has been legitimate for centuries will now be burdened by layers of bureaucracy and red tape, which will catalyse corruption. India's thriving buffalo meat exports, which recorded ₹26,684 crore in 2015-16, will be crippled by the move. Milch cattle have a limited productive period. The livestock economy therefore depends on the farmer's ability to sell cattle subsequently. Other industries such as meat, leather, soap, automobile grease among others, use livestock or its by-products as inputs. Therefore, restricting livestock trade impairs other industries and threatens millions of jobs.

Moreover, article 246(3) of the constitution gives states exclusive powers to legislate on preservation of cattle. By encroaching on this domain, we are headed towards greater friction and contestation between Centre and states. To make matters worse, vigilante groups which have functioned unchecked may choke even the limited livestock trade permitted by the new rule. Irrationally the Law promotes overall levels of irrationality in society.

The new rule will not be effective right away. Therefore, the government should annul it before adverse economic and political consequences show up. The argument that the rules aim to prevent cruelty to animals is unconvincing - they will in fact promote cruelty to animals when farmers are forced to resort to underhand means to dispose of unproductive cattle.

The Indian economy today produces few jobs relative to the number of young people coming into labour markets. It is, therefore, imperative that the government need not to introduce rules that destroy even jobs that exist currently.

Impact of Ban on Cow Slaughter in Maharashtra: Maharashtra's decision to extend the ban on cow slaughter to bullocks in March 2015 cut off the demand from abattoirs which were key customers for old livestock. Now the ban has virtually frozen it. Old bullocks that used to draw around ₹20,000 a pair are not selling. Prices even for younger livestock have halved from ₹60,000 a pair to ₹30,000. Prime cattle which rated upto ₹1 lakh is down to ₹50,000. Why would anyone buy bullocks if they cannot resell them when they are old (Source: HT 05.06.2015).

Impact of Ban on Meat Trade in UP: Not only have the legal slaughterhouses and meat shops been closed but a large number of abattoirs, otherwise legal, have also been sealed by the administration in UP for violating one or the other rule. 140 of the 185 legal slaughter houses along with hundreds of illegal ones have been closed down since the government's statewide crackdown began on March 21, 2017. Thousands of meat shops also shut down either voluntarily in solidarity with other slaughter houses or due to a shortage of meat supply or because they simply did not comply with the terms and conditions of the license issued to them.

Cow Protectin Laws Affect Mostly Muslims and Dalits: Bearing the brunt of the crackdown are obviously Muslims because it is mostly they who are into the meat business. Thousands of Muslim families that sustain on meat trade in the city are facing a very difficult time. 19% of the country's meat production is in Uttar Pradesh followed by Andhra Pradesh and West Bengal. 65 of the country's 109 big export oriented integrated abattoirs and meat processing units are in UP. Aligarh, Ghaziabad, Meerut, Bulandshahar, Moradabad and Unnao are the main centres.

Cow protection laws in force in as many as 24 states across the country also infringe on Dalit rights. Dalits care for cattle when they are alive, they also skin dead animals and dispose of the carcasses. With these laws, the leather industry, cattle rearing and skinning of dead animals all get affected.

Impact of Ban on Cattle Smuggling Across the Indo-Bangla Border: Every year close to 25 lakh cattle are smuggled from India to Bangladesh, much of which happens with the connivance of those guarding the border. Besides massive demand in Bangladesh, there is matching supply from India as dairies and independent cattle-rearers can't afford to feed their animals after they have gone dry. With 25 lakh cattle approaching borders every year, if the cattle smuggling is halted it would be saddled with 1.25 crore dry cattle to feed annually. Given the cumulative cost of sustaining a dry cow for a year would translate to roughly ₹25,000, the government would have to shell out ₹31,250 crore annually (Source : TOI-04.04.2015). A cow that costs ₹5,000 here could fetch upto five times that price once it crosses the border. And handlers could earn anywhere between two thousand and ₹5,000 for smuggling out a single cow. That is why this business is so rampant here.

While there were enough cows in Bangladesh to satiate the local need for beef, the cows that were smuggled out from India were primarily intended for the middle east market. The cows that are smuggled out from here are not slaughtered. They are killed slowly in boiling water and their hide is peeled off. Then the meat is sliced and packaged as various products, such as beef sausages, ground beef and even steak slices, before being exported. Beef exports represent the biggest industry in Bangladesh. Healthy cows, bred under the right conditions, can fetch upto ₹100,000 each. According to a Reuters report, at least two million heads of cattle were smuggled into Bangladesh from India every year. Annually, this trade was worth at least US$600 million. The smuggled cows were auctioned by Bangladeshi traders to facilitate the sale of cattle to slaughter houses, beef-processing units, tanneries and bone-crushing factories, which contributed approximately 3 percent to the coutnry's $190 billion economy. Most of the beef flesh and beef products were marked for export and sent to the Middle East; Singapore, Malaysia and other countries in the Far East. All the beef consumed in the Arab countries and in South East Asia are processed from Indian cows.

India is home to 300 million cattle and is the world's largest exporter of beef. It exported $4.8 billion worth of beef in 2016 and it is fifth-biggest consumer. Recently, the rhetoric of cow protection and beef banning has spun out of control. In March, the Maharashtra Government extended the ban on cow slaughter to include bulls and bullocks. A few weeks later, the state of Haryana passed similar legislation.

Legal Diversity about Cow Slaughter

Legal diversity in governing cattle slaughter in different states are mentioned below (Status as on 12 October, 2014).

- There is lack of uniformity among state laws governing cattle slaughter.
- Cattle slaughter is fully or partially banned in 24 states.
- There is no ban in Kerala, Tripura, Sikkim, Arunachal Pradesh, Manipur, Meghalya, Mizoram and Nagaland.
- The strictest laws are in Delhi, Gujarat, Madhya Pradesh, Punjab, Rajasthan and Uttar Pradesh.
- Assam, Tamil Nadu and West Bengal permit slaughter of cows above a certain age.
- In Delhi, Goa, Puducherry, Punjab and UP, cattle slaughter is a cognizable and non-bailable offence.

- The maximum term of imprisonment in states varies from six months to 10 years.
- It is difficult for restaurants to source or serve beef legally but beef consumption is prevalent all over India.

Norms of Slaughterhouses in UP (Source: TOI-02.04.2017)

- Stay outside a 100-metre radius of religious places.
- Stay away from vegetable markets.
- Get NOC from civic body, circle officer, FSDA in cities.
- Get NOC from gram panchayat, circle officer and FSDA in villages.
- Any FSDA violation will result in immediate suspension of licence.
- All meat shop workers must get health certificate from government doctors.
- Get meat certified from veterinarian.
- Shops must have curtains or tinted glass to ensure meat is not visible to public.
- Can't slaughter ill, milch or pregnant animal.
- Must get premises whitewashed every six months.
- All knives and blades in shop must be of steel.
- Must transport meat only in insulated freezer vehicles from slaughterhouses.
- Must have complete records of slaughterhouse purchases.
- Should have proper disposal of waste.
- Should have refrigerator with transparent doors.
- Should have a geyser.

Bill to Protect Cow Slaughter in Gujarat: The Gujarat assembly on 31 March, 2017 passed a bill enhancing punishment for cow slaughter from the present 7-year jail term to life imprisonment. It includes a fine of upto ₹1 lakh and permanent confiscation of vehicles found being used for transportation of cows illegally. The offence is also likely to be made cognizable and non-bailable (Source: TOI-01.04.2017).

Cow Crimes and Punishment in Haryana: Haryana enacted one of the most stringent laws in the country against cow slaughter. The guidelines of the Haryana Gauvansh Sanrakshan and Gausamvardhan Act 2015 are mentioned below.

Cow Crimes and Punishment in Haryana

Offence	Punishment
Cow Slaughter	Jail -3-10 years, fine ₹30,000 to ₹1 lakh
Export of cows for slaughter	Jail 3-7 years, fine ₹30,000 to 70,000
Sale of beef	Jail 3 years to 5 years, fine ₹30,000 to ₹50,000

(Source: TOI-26.10.2016)

3

Cow – a Moving Wealth

India has a large cattle wealth. It plays an important role in the socio-economic life of people in general and in the villages in particular. It provides vast opportunity for self-employment. About 20 million people are engaged fully or partly in India. This sector contributed 6.5 per cent to the exchequer from 1949-2000. It is the main or supplementary source of income of the weaker sections. It is also a source of raw material for several manufacturing industries. There are states such as Rajasthan where cattle are the only wealth of the people. Why and how a cow a moving wealth is narrated below (Source: HT-01.06.2017).

- It is believed that 33 crore Gods and Goddesses reside inside the cow and the animal appeared along with goddess Laxmi during the churning of ocean in Hindu mythology.
- Cow is the only living being which intakes oxygen and emits oxygen and the animal is a hospital in itself.
- Cow urine keeps the liver, heart and mind healthy, and increases the immunity of the body. It also slows down ageing.

- Drinking cow urine rids one of the sins of the previous life.
- According to a Russian scientist, coating the walls of the house with cow dung will protect the inhabitants from radiation.
- The cow, through its horns, absorbs "cosmic" energy.
- Consumption of cow milk prevents cancer from entering the blood cells.
- Cow dung known to have killed cholera germs.
- Cow dung can generate 4,500 litres of biogas every year. If biogas is generated from the cow dung of all the cow progeny in the country, the country can save 6.80 lakh tones of firewood and save 14 crore trees from being cut.
- Cow is the only mammal whose large intestine is 180 feet long because of which the cow milk has Kerotin which produces vitamin A in human body.
- Mooing of cow kills the pathogen in the air.

Chemical fertilisers are injurious to the soil. They stimulate the soil and make it produce much for the time being, but leave it in the end exhausted and impoverished. They are also said to cause diseases in crops and in animals. Only manure which can permanently enrich the soil and help healthy growth in plants and animals is organic that includes the cattle dung and urine, human excreta, and waste vegetable matter. This makes cattle relevant in agriculture, not only for labour but also for the valuable manure they provide. Cow based products are bio-fertilisers, bio-pesticides, bio-energy and panch-gavya.

A Fuel, manure, plastering, smoke repellent for mosquotes, ash for preserving grain and utensil cleaning and medicinal use. Dung from one cow fulfils cooking and heating requirement of a 5-member family in contrast 10 trees (5 years old) are required per year to meet the requirement.

The contribution of the cow through its milk, bullocks, manure, hide and bone to the wealth of India is over ₹1,000/- crore annually, an amount which no other industry in India except agriculture can be equal.

Around 200 years ago, there were some regions of the country where the yields of agriculture produce were more than what it is now in Japan. This was mainly due to utilisation of cow like cow dung and cow urine as manure/pesticides and use of bullocks for ploughing the fields.

Energy from Cattle

Generates more energy than all of India's generating plants- 800 MT manure/year. Its replacement with coal will cost India $1.5 billion/year. Shifting Indian agriculture from bullocks to machinery has costed India $ 30 billion plus maintenance & replacement cost extra.

Bullocks Suit to Our Agriculture System

India is an agricultural country. About 300 millions of its population depend on agriculture for their livelihood. For them the cow is more than their right hand, since without the aid of bullocks which the cow provides, ploughing, irrigating, weeding, harvesting, threshing, carting and marketing will be next to impossible. Bullocks are also necessary for carrying on village industries like oil-pressing, At present these animals, which provide the motive power in agriculture and village industries, are weak, starved and diseased. How can people in our villages become prosperous so long as they have to depend on such a feeble instrument for eking out a livelihood?

Cultivation through the bullocks maintains soil condition which is very delicate and thin. On the farm, bulls are used to plough the fields and as a means of transportation of goods. Even Lord Shiva's trusted vehicle is

Nandi- the sacred bull. The cultivation through the bullocks is most suitable for maintaining soil conditions. The invasion of tractor and chemical fertiliser wiped of the cows and bullocks and made them 'useless' from 'useful'.

Even later, it may not be wise for our farmers to adopt tractors and mechanical devices in agriculture. They are useful in countries with a small population and vast areas to be cultivated. The situation in India is just the reverse. The peasant at present has on an average one or two acres in our county. If we adopt machines to replace human beings in agriculture, where shall our people go for employment? Large scale industries are not able to absorb more than about two million. If agriculture also is mechanised, it too will not be able to provide work except for a few millions. What will happen to about 400 million (40%) people who are engaged in agriculture and allied sectors. For these reasons, then it would seem best for us not to be allured into following the way of mechanised agriculture but to fall back on the bullock for motive power. If we do so, the cow which provides the bullock must occupy a central place in our national economy.

How to Save Bullocks

Before mechanisation, agriculture thrives on bullock power. Bullock carts were used to move majority of the goods across the nation and used to do two-thirds of the work on the average farm. Converting Indian agriculture from bullocks to machinery has costed to the country an estimated thirty billion dollars plus maintenance and replacement costs. Cross-bred bullocks are not able to keep up the pace with the zebus. Inspite of massive mechanisation of farm operations, most of the agricultural operations and transportation in rural India particularly backward regions of the country still depend to a large extent on animal's power. Machinery may be introduced when the existing man and animal power is fully occupied. At present this power is not fully utilised, and therefore there is no reason for the introduction of machinery.

Now that mechanisation threatens to overrun agriculture and transport as a part of our planning. Therefore, need is to sum up the case for the bullock which is doomed to destruction. We must have milk, more and still more milk. We must therefore have cows, and if we have cows, the bullocks will always be with us, for these we have to provide and can provide full employment only if we yoke them to the plough, to the cart and to the ghani. The main agricultural operations of ploughing, harrowing, sowing and inter-culturing keep the bullocks busy for only three or four months in the year. During the rest of the year they should be used for carrying goods as well as passengers, for crushing oilseeds and so on. If we fail to do this, we shall be reduced to the same plight as the Western nations who slaughter all bull-calves except a few which are reared as stud bulls. The bullock is not only a living tractor, it is also a living fertiliser factory and gives us farmyard manure which supplies nitrogen and improves the porosity of the soil, thus helping to increase the moisture content of the soil as well as proper aeration. These three factors are essential to plant growth. No amount of concentrated manure would help if the porosity of the soil and consequent aeration of the soil are not improved.

Cow-based Rural Economy

Although, rural development has been the focus of various government schemes and policies since Independence, one of the ways is giving enough importance to the cow-based rural economy. Cow-based rural economy means the use of five key products from cow called Panchgavya — milk, curd, ghee, dung and urine. These have been a part of daily life in the Indian Subcontinent in the form of food, medicine, agriculture, etc. Concept of organic farming and increased demand for cow based products such as bio-fertilisers, bio-pesticides, bio-energy and panch-gavya medicines gives an opportunity to make agriculture economically viable on a sustainable basis.

Panchagavya

Panchagavya or panchakavyam is a concoction prepared by mixing five products of cow and used in traditional Indian rituals. The three direct constituents are cow dung, urine, and milk; the two derived products are curd and ghee. These are mixed in proper ratio and then allowed to ferment. The Sanskrit word Panchagavya means "mixture of five cow products".

Uses of Panchgavya: It is called cowpathy treatment based on products obtained from cows used in Ayurvedic medicine and of religious significance for Hindus. A common usage is as a fertiliser and pesticide. Seeds can be treated with panchagavya. This was found useful in rhizome of turmeric, ginger and sugarcane and they yielded more. It helps in plant growth and improves immunity.

All pathies such as Naturopathy, Unani, Siddha and Homoeopathy cite the benefits of Panchgavya: such as 1) Strengthens immunity, 2) Aids digestion and detoxifies digestive system, 3) Has anti-bacterial and fungicidal properties, 4) Helps relieve symptoms of a range of eye and skin conditions, 5) Is a good natural exfoliator and 6) Contains antioxidants.

The five subject areas of Panchgavya are uniqueness of indigenous cows, Panchvagya in agriculture, Panchgavya in Medicine and Health, Panchgavya in Food and Nutrition and Panchgavya for Utilities, Products and Processes. Companies such as Vedic Cow Products, Calves N Leaves, the Thela and Patanjali Yogpeeth are selling a range of products infused with cow urine, dung, milk and milk products. These include Gau Ark (distilled cow urine); Soap; Nasal Drops; Face Pack; Hair Tonic etc.

Several institutes/organisations are working to validate the health benefits of Panchgavya - a concoction prepared with cow urine, dung, milk, curd and ghee is mentioned below:

- Studying benefits of urine of indigenous cows in improving immunity and treatment of cancer.
- Investigating cow urine for increasing efficacy of tuberculosis treatment.
- Finding antiviral potential in cow products to treat chikungunya and dengue.
- Developing probiotic drink and insect repellants from cow products.
- Establishing 'Gau Vigyan University' as a centre for all research on cow products.

Panchgavya Research Panel: The government has set up a 19-member panel to carry out scientific validation of research on cow-derivatives including its urine, and their benefits. Headed by science and technology minister Harsh Vardhan, the committee will select projects that can help scientifically validate the benefits of panchgavya - the concoction of cow dung, cow urine, milk, curd and ghee - in various spheres such as nutrition, health and agriculture. Named the National Steering Committee, the panel includes secretaries of the departments of science and technology, biotechnology, Ministry of New and Renewable Energy, and scientists from IIT Delhi. It also has three members of Vigyan Bharti and Go Vigyan Anusandhan Kendra. The government has given the project to SVAROP (Scientific Validation and Research on Panchagavya). It will cover "scientific validation of panchagavya for medicines and health, scientific validation of panchagavya and its products for agriculture applications, scientific validation of panchagavya for food and nutrition and scientific validation of panchagavya based utility product. The panel, which will have a tenure of three years, will function as an apex body for guiding SVAROP.

Cowpathy

A New Version in Medical Science: Cowpathy is a system of medicine where-in, the human or animal ailments are treated with products of cow also known as Panchgavya means five things derived from cows i.e. milk, ghee, curd, urine and dung. In our ancient literature of Ayurveda it is described as Panchgavya Chikitsa. As modern allopathic system is facing problem in handling the multiple drug resistance in microorganisms, presence of antibiotic residues in food chain and/or associated allergies and autoimmune disorders, the alternative system of medicine is gaining momentum not only in India but abroad as well.

During the last few years it has been observed that the efficiency of 'Phagocytic System' through macrophages (monocytes of the blood) reduced drastically as a result of the environmental pollution and presence of pesticides, heavy metals, fungal toxins etc. in the food chain. That is attributed to the heavy use of agrochemicals in agriculture and poor storage conditions of food grains. Anyhow deficient functioning of macrophages leads to inefficacy of antibiotic drugs, development of resistance in bacteria, recurrent infections and /or decreased immune status of an individual. Recent researches showed that cow urine enhances the immune status of an individual through activating the macrophages and augmenting their engulfment power as well as bactericidal activity. This research opened a new era in medical science and India has got a patent from US on cow urine for its bioenhancing properties and its use in tuberculous patients.

Very recently, researches have found that use of cow urine along with the antibiotics prevents the development of resistance in microorganisms against the antibiotics. The cow urine blocks the R-Factor, a part of plasmid genome of bacteria responsible for the development of anti-bacterial resistance. Several scientists from different laboratories including G.B. Pant University, Pantnagar, and IVRI, besides NGOs are working on different medicinal properties of cow urine. In fact there are several medicinal preparations available with NGOs who are also marketing cowpathy drugs under FDA license and some of the NGOs are even not able to cope up with the demand of the public. Cow urine and other products of cows have been found to enhance the immunity and resistance power to fight with infections. Cow urine has antioxidant properties and thus it neutralises the oxidative stress produced in body through action on free radicals. It has been found to repair the damaged DNA and thus is effective for the cancer therapy.

There are several NGOs working in Kanpur, Nagpur, Jaipur, Rishikesh, Haridwar and other parts of the country who are in this business. Mostly they prepare Ark of cow urine i.e. 60-70% distillation of cow urine. But, on the basis of chemical fingerprinting through modern equipments like HPLC, it has also been observed during the scientific research that the urine of Indian cows are highly effective and almost nil or few medicinal properties are present in the urine of crossbred, exotic cows, buffaloes, etc. The Indigenous cow urine contains "Rasayan" tatva, which is responsible to modulate immune system and acts as bio-enhancer. It is not only the cow urine which is a wonder product by cow but other products like dung, milk, ghee and curd are also equally effective for various ailments and other operations. Cow urine as such and/or after addition of neem leaves is a wonderful bio pest repellent. Such bio pest repellents are safe to use, do not accumulate in the food chain and as such do not have the harmful effects like chemical pesticides. Cow dung is an excellent farm yard manure and if processed and prepared vermicompost, a very small amount of vermicompost is sufficient for a large field. Similarly, many medicines are prepared from the cow milk, ghee and curd. However, scientific validation of these products is required.

Frustrated from the heavy medication of allopathy people are now using cowpathy drugs.

In the time to come, the field of cowpathy will come up with strong scientific facts and hopefully dominate on other systems of medicine. Hence, this old system of Panchgavya medicine at present known as cowpathy may get popularity and provide relief to the ailing people and give them life without pain (Source: ICAR News October-December 2017).

Dairy Products

Milk is white gold. Its products constitute about 35% of the total packaged food industry in India. About 37% of the milk sold is used for processing, with the organised dairy industry accounting for 20%, and the unorganised sector for 22%. Dahi accounts for ₹15,000 crore per annum, with the organised sector accounting for 10%. The organised cheese market is pegged at ₹5,000 crore.

The demand for processed dairy products through supermarkets is expanding and is expected to keep doing so as cold supply chains and modern trade continue to evolve. Increasing urbanisation, exposure to gourmet foods and corresponding changes in consumer preferences, behaviour and purchasing power are the catalysts for the rise of processed dairy categories. Why processed dairy foods will grow, the reasons are mentioned below (Source: HT 23.05.2011).

(1) Supermarket sales will expand at a much higher rate than other retail outlets. Processed, value-added, packaged dairy products will find place in these locations.

(2) The processed food market, including dairy, is attracting foreign companies due to high potential for growth, as they are underdeveloped.

(3) The growth of modern, organised retailing will also increase the demand for value-added foods, including dairy products.

(4) Rising household income, increasing urbanisation, changing lifestyles and the rapid growth of the private-sector and dairy-processing industry will lead to greater demand for value-added, milk based products.

Cow Milk: Cow milk has a prominent place in the diet of people, from an infant to an elderly person. It is compared to the mother's milk for the quality for kids. A2 milk produced by indigenous cows prevents disorders like obesity, arthritis, type-1 diabetes among children, autism, etc. Substantial research has been done worldwide highlighting such medicinal significance of A2 milk produced by indigenous cows. A2 milk contains A2 type of beta casein protein rather than more common A1 protein found in common milk. It is characterised by a distinct golden colour. A2 milk has 12% more protein, 33% more vitamin D, 25% more vitamin A and 15% more calcium. By drinking A2 milk one is no longer exposed to BCM-7 found in A1 milk which is linked to neurological impairment, type-1 diabetes, impaired immune response, autoimmune disease and heart disease. A2 milk is easily digestible and has good quality fat.

A1 and A2 cow milk has mainly two types of proteins viz., casein and whey protein in 80% and 14% proportions. Among casein in milk, β-casein is the second most abundant protein comprising 33% of protein in cow's milk. It is important because of its structure and stability of casein micelles and has an excellent natural balance of amino acids. Out of the 12 known variants of bovine beta casein, A1 and A2 are most common. Cosmophin (BCM-7), product of a variant of A1 β-casein gene, has been implicated in a number of risk factors to human health, viz., type 1 diabetes mellitus, obesity, arteriosclerosis, sudden infant death syndrome, neurological defects including autism and schizophrenia. About 90% of animals tested in India were found to belong to A2 A2, 9.1% as A2 A1 and only 1.0% as A1 A1. The frequency of A2 A2 was 76%, 40% and 33% in indigenous, crossbreds and holsteins, respectively. None of the indigenous breed bull was homozygous for A1. New Zealand has now established A2 Corporation Limited to market A2 milk at a premium price throughout New Zealand and Australia (Source: Agric Today - Sept. 15, 2017).

Cow Ghee: Cow ghee from indigenous cows has been used since time immemorial for its benefits. Other reasons which may be given in favour of the cow for supply of milk as against the buffalo are (a) that cow's milk is more conducive to health than buffalo's as it has more vitamin B, and has in addition vitamin E which is absent in buffalo milk, (b) that the carotene (vitamin A) value of cow's ghee is ten times as high as that of buffalo ghee.

Curd: Curd and buttermilk have been found useful in many gastrointestinal disorders and are recommended as a food practice in Ayurveda.

Cow Dung: India is the largest cow dung producer worldwide. In parallel to the milk industry, also the occurrence of dung will further grow from an estimated current total output of 2 million tones of dung per day to over 3 million tones in 2022. A dairy cow produces about 115 pounds of manure each day, carrying on an average 10 pounds Nitrogen, 4 pounds Phosphorus and 8 pounds Potassium per tone of manure. Although nutrient content per pound may seem small, still this is equivalent to 200 to 600 pounds of nitrogen per acre (Source: Agric-Today-July, 2017).

According to Ayurveda, Gomeya/ (cow dung) is not a waste product but it is a purifier of all wastes in the nature. When spread over urban and rural waste in solution form (1:10-1:25 solution), it biodegrades the waste in no time. It is a 'Gold mine' due to its wide applications in the field of agriculture, energy resource, environmental protection and therapeutic applications. Cow dung obtained from Indian Indigenous cow/Bos Indicus/Zebu breed is better than that of exotic breeds. Ideally, the source of cow dung as per Ayurveda should be from a healthy Zebu cow.

Dung-economy is in fact the nucleus of prosperity of the Indian people. India's cattle produce 800 million tones of manure every year. Most of the dung is used as fertiliser or fossil fuel. Cow dung is high in methane, and can generate heat and electricity. Many village homes are plastered with a mud/cow dung mixture, which insulates the walls and floors from extreme hot and cold temperatures. Cow dung being rich in minerals, makes an excellent fertiliser. There is a big organic farming movement in India to return to ancient methods of utilizing cow dung to re-mineralise the depleted soil. Cows also play a central role in fire yagnas or Agnihotras. Scientific research has found that the ritual of burning cow dung and ghee as fuel for these sacred fires, actually purifies the air, and has anti-pollutant and anti-radiation qualities in the environment.

Cow dung is also the cheapest and most easily available fuel to the rural families in the country. Cooking and heating requirement of a villager having a family of 5 members, has been calculated to be around 1.25 million Kcal/year and this requirement is mostly met from dung. Dung cake obtained from one cattle is sufficient for a family for a year. While burning these cakes, the temperature never rises beyond a certain point; ensuring overheating does not destroy the nutrients in the food. The use of cow-dung in biogas as a non-fossil fuel is being considered for vehicles and cooking. Cow dung also has antiseptic, anti-radioactive and anti-thermal properties. Even ash from cow dung has several uses including preservation of food grains. In olden days, the ash from cow dung was used to preserve food grains as buffer during drought year. Food grains equally mixed with ash could be stored for 12 years without any damage. Even now the people in villages used ash to store food grains for 2 to 3 years in their storage tanks made of clay. Cleaning of utensils was another important use of ash and was one of the routine and essential household chores, for centuries.

Gobar-Dhan Scheme: The scheme roadmap is not just meant as a means to convert waste to wealth, but is an ambitious effort to clean India's villages. According to estimates, India has about 30 crore cattle population and approximately 30 lakh tones of cattle dung is generated daily. Other biomass such as crop waste and kitchen waste are fantastic source to generate bio-CNG and manure. The scheme will be entrepreneur-led and ensure the bio-CNG and manure are sold to make the programme sustainable.

To begin with, the drinking water and sanitation ministry will roll out the scheme across 350 districts from April 2018 and in the second phase rest of the districts will be covered. One of the major oil manufacturing PSU has agreed to buy the bio-CNG that will be produced under this scheme. The scheme will involve bio-mitras, who will collect the waste and will get remuneration. The local panchayat will be the centre of the scheme. They will take the decision about technology and where to set up the plant. The government will help the entrepreneurs under this scheme and it will be output based.

Even after her death, the cow's horns and other parts of the body are used to make manure which is very rich in nutrients for the soil. Replacing dung with coal would cost India $1.5 billion per year. Cow dung is also used for plastering the outside walls of a house which keeps the house warm in winter and cool in summer. Cow dung if burnt, its smoke acts as a repellent for mosquitoes. To prevent soil from becoming barren, it is necessary to apply 10 bullock-cart loads or 5 tones of dung manure for each acre of land.

Cow Urine: Cow urine helps in anger control. It helps attain a 'cow-like calm nature' and can prevent people from committing crimes like assault and murder; Can help keep ghosts at bay; Can help reduce weight; Urine has gold, copper, silver, manganese and iron. Horns of a cow attract cosmic energy which seeps into its milk, urine and dung (Remedies from Aarogya Geeta on Gujarat Gauseva and Gauchar Vikas Board website).

Cow Urine also has the potential in treatment of bacterial infections and cancer, and demonstrate that cow urine can enhance the efficacy and potency of other drugs. Two US patents on cow urine (No. 6896907 and 6410059) have been granted for its medicinal properties, particularly as a bio-enhancer and as an antibiotic, antifungal and anticancer agent.

Uses of Cow Parts: Cow parts are used for different purposes: 1) Edible fats used for margarine, chewing gum and certain candies, 2) Gelatin from bones and skins used for marshmellows, ice cream, desserts, 3) Intestines for sausage casings, 4) Inedible fats for soaps, lipsticks, face creams, certain medicines and ingredients for explosives, 5) Bones, horns and hooves for buttons, bone china, piano keys, glues, fertilisers, photographic film, paper, wall paper, sand paper, combs, toothbrushes, and violin string and 6) Cow's body parts said to be used in more than 100 drugs.

4

Cow Menace

Sadly, over a thousand years of rule by foreign rulers and religions has meant the once "sacred cows" of India are now as destitute as much of its ancient Hindu culture. More often, you will find our cows roaming on the roads of India, sitting idly in the middle of a traffic jam, chewing cud of the nearest waste heap and a plastic bag tangled up in the horns. This is a truly sad state of affair. Not because the cow is sacred, nor because it has lost its place in the list of what we consider as precious. But, because this is not how we should treat an animal, may be cow or other.

Earlier, at no time whatsoever, would a cow be allowed to wander the streets unsupervised or without its owner. The only time cows were seen in the streets, was when they left to go to the pastures in the morning and when they came back. At all other times, they were either in the pastures or in the cowshed. Why would you let your most precious possession roam around at will? Would you leave your Mercedes with the keys in the ignition switch and windows rolled down? To the ancients, our current attitude towards cows would be equally incomprehensible.

Cows are subjected to daily oxytocin injections causing excruciating labour-like pains, in the hope of increasing the milk flow. Even calves are deprived of their share of milk. Newly born male calves are largely unwanted. They are sent to slaughter, or sent out to fend for themselves. Or they are tied to a pole allowed to die slowly of starvation.

Late Shri Nathuram Mirdha, the then Chairman of the National Agricultural Commission, when asked by a foreign delegation as to what, according to him, was the culture of India. His reply was "Agriculture and no other culture is the culture of India" and the king-pin of agriculture in our country is the cow, other-wise known as "Kamdhenu". Now situation is fast changing. Not only productivity is poor, there has been negative growth rate in cattle population in nineties. The cow whom we call "Gaumata" is becoming a burden on the society. Reasons are explained below.

Blind Following of Western Agriculture

Blind following of western agriculture, particularly in respect of mechanisation of agriculture without looking into the nature of our soil, size of our holdings, number of man power available, the capacity of our bullocks and ignoring completely our traditional methods of farming has created havoc to our farmers as well as to agricultural labourers. Use of chemical fertilisers and pesticides on a massive scale has not only increased the cost of production but has also poisoned our soil, water and crops and caused health hazards to the people and cattle.

Indiscriminate Breeding

In order to improve the productivity of our cows, they are being indiscriminately cross-bred. The exotic and cross- bred cows and calves are misfit to our climate and unable to tolerate the heat of summer. Exotic animals brought with them many diseases not existed earlier in our country. The quality of their milk is inferior to that of the indigenous breeds. Their male calves have no humps and are not capable of ploughing etc., as are our indigenous breeds. Indiscriminate breeding has led to dilution of our precious indigenous breeds and few cow breeds are at the verge of extinction.

Illegal Slaughtering

A report by the Union Ministry of Environment (MOE) claims that the massive illegal slaughtering of animals has posed a serious threat to their existence and there has been an alarming decrease of animal - human population ratio. The report also states that the cattle human ratio came down from 430 animals per thousand human population in 1951 to 278 in 1981 which has now reached the alarming level of 200 per 1000. Central Leather Research Institute, Chennai in its analysis mentioned that "indiscriminate slaughter of animals to meet the internal and export demand for meat had contributed to the low growth rate of animals. Cattle population, which was 204.58 million in the year 1992 decreased to 195.58 million in 1997. For the first time since 1951, a negative growth in the cattle population of the country has been noticed which is indeed a cause of great concern and sound warning bells for cattle wealth.

Feed and Fodder Deficiency

Livestock production particularly ruminants in India depends on crop residues and crop by-products. There is a wide gap in demand and supply of animal feed and fodder. The overall deficiency of feeds and fodders is estimated to be about 30 percent in respect of total digestible nutrients (TDN) and as high as 70 percent in digestible crude protein. Under feeding, protein and vitamin deficiencies and imbalance of trace elements result in various reproductive abnormalities both in male and female. Poor nutrition during early life in female retards the onset of puberty. In parous cows, it leads to anoestrus, anovulation and metabolic disorders. During early pregnancy, moderate to severe under-nutrition may lead to embryonic mortality and/or abortion. During late pregnancy it has an adverse effect on calf birth weight. A poor body condition score (BCS) at calving adversely affects milk production and fertility characterised by prolonged post-partum oestrus interval, reduced conception rate and more services per conception. The nutrient deficiencies and imbalances during initial period of lactation can lead to several disorders, which affect the post-partum reproduction. A very low protein diet can cause cessation of oestrus; if fertilisation occurs, foetal death or the birth of premature and weak offsprings result. The plane of nutrition also influence the performance of breeding males. Ration containing low protein adversely affects semen volume and quality. Feeding of low protein ration for long also affects sex libido.

At present situation is pathetic. Cattle are left to roam and find food themselves. They feed on roadside garbage supplemented with some feed by their owners. Many die prematurely due to their stomachs obstructed by plastic bags. Recently doctors removed 80 kg of polythene waste from the stomach of a cow at Patna, Bihar veterinary college on 14 February 2018. How pathetic it is for an animal whom we call Gaumata. It is shame to the country and to the vigilantes who claim themselves Gau Rakshaks. Bullocks often collapse while transporting large loads in the hot sun without rest, food and water.

Poor Reproductive Management

Breeding efficiency is determined by many different processes that include age of puberty, pattern of oestrous cycle and oestrous behavior, length of breeding season, ovulation rate, lactational anoestrus period, inter-calving period and reproductive lifespan. Heretability of breeding efficiency is 0.30. Selection based on this trait is advantageous in terms of reduction in age at first calving and first service period. Ample scope exists for increasing the reproductive efficiency by modification in the traditional methods of breeding, feeding, management and disease control.

Adverse Climatic Conditions

Environment is one of the determining factors in the production and reproduction of farm animals all over the world. Inborn production potentialities can be severely affected by exposure to adverse climatic conditions. Season affects the breeding efficiency in animals. Season of calving affects lactation performance, lactation length and calving interval. Winter calvers produce more milk in a lactation as compared to those calve in rainy season. Those who calve in rainy season have least calving interval. Oestrogen activity of some fodders also influences seasonal activity. Measures to ameliorate climatic stress help in improving reproductive efficiency. Such measures are very important in exotic animals.

Lack of Forces of Selection: Other factor responsible for poor production is lack of forces of selection as a result of which the poor yielders do not get out of the population and hence contribute off springs from generation to generation. Slaughter is being used throughout the world to restrict an increase in population of undesirable livestock. The need for slaughter is also important if there is a programme of genetic transformation of the population through crossbreeding. In view of the religious taboo and legislative restrictions on the slaughter of the cows in many states of India, the poor yielders do not get out of the population, and contribute off springs from generation to generation. These low yielding cows compete with better stock in respect of feeding, housing and other managerial necessities.

Lack of Use of Modern Breeding Tools: Advances in animal breeding during the past few decades have been phenomenal.

Farm animal breeding has entered into an era of new biotechnology revoluation which includes artificial insemination, embryo transfer, cryopreservaion, sexing, transgenesis, cloning etc. These technologies still in infantile stage in India, have revolutionised animal agriculture in developed countries of the world.

Why Are Cows on Indian Streets?

Cows are worshiped in India and feeding cows is often considered to be an act of holiness and goodness towards animals by most of the Indians. Now, the situation has changed. The milkmen in India think, let their cows roam free as they know that they will get free food. The fodder which they feed to the cattle comes moderately expensive. The cows move from one house to the other (as there is scarcity of green lush grass field in cities) to get feed by the people living in the houses. It cuts down the fodder expenses to a great extent for the cow owners. It's like double jackpot for them, because the cows return to their owner's place to give milk after eating its food. The milkmen sell the milk the cow produces, eating all the free food plus the fodder fed by the owner. So, more milk for the owner as well as the calf. It's a smart business idea. It's not a bad thing. After all, feeding cows, or any other animal for that matter, according to the Hindu religion, is believed to be an act of goodness and humanity. Another reason for cows roaming freely on roads is mismanagement of the local governing bodies. The people responsible for keeping the cows away from the roads are sometimes bribed by cow owners, so that the cows don't get captured by them.

For most Indians, though, these creatures are n't just religious symbols — they're neighbors. In New Delhi, for example, it's not unusual to find a cow loitering, unperturbed, in the middle of the road. Unafraid of traffic, they often cause logjams or accidents. Commuters, however frustrated by the delay, can do nothing but try to convince the cow to move along, because hurting one in any way would enrage Hindus.

There are 191 million (68 male and 123 female) cattle in India. Number of stray cattle is 53 million (livestock census 2012). Hundreds of cattle on the roads feed on garbage, including plastic. They are sentenced to a slow and cruel death if they do not receive help in time. Big cities like Delhi, Kolkata, Chennai and Bengaluru where tens of thousands of animals roam with bellies full of plastic.

The problem of stray cattle on roads is not new in India, and these animals often bring the flow of traffic on busy roads to a standstill along with major road accidents. The situation turns all the more worse at night when it becomes difficult for the drivers to spot the stray animals due to darkness. The police are also silent over the issue. These abandoned animals can be spotted at many points in the city.

Most cows roaming in the street of cities are not healthy cows. They do not live on grass or other hygienic foods. Most of these cows constantly eat food from garbage and leftovers. Many of them are actually sick. The problem of stray animals, causing traffic blockades and accidents on city roads, is increasing day-by-day due to the lackadaisical attitude of the administration towards the issue.

These animals are reservoirs for several of the most important food-borne and waterborne agents. Manure and wastewater from animal feeding operations have the potential to contribute pathogens and pollutants, such as antibiotics, nutrients (nitrogen and phosphorus), hormones, sediments, heavy metals, organic matter, and ammonia, to the environment and eventually could impact human health. Some of the illnesses may cause severe crises and even death. Many lives have already been lost and many others have been injured due to accidents involving stray cattle, especially in rainy season, due to the cow dung roads becomes slippery which invites the road accident in great way.

Where Do They Come From?

Bulls, the male calves and cows that have "dried out" are often turned out of dairy farms because they do not produce milk. Some farmers sell such cattle to slaughterhouses, but many others are discouraged by religious objections and the militant Hindu "cow protection" groups, who have taken to attacking trucks that transport cattle and beating up farmers who they suspect of sending cows to their death. Abandoning cows is often considered the safer option. Even cows that are n't abandoned often find themselves on the streets — illegal dairies cut costs by allowing the cows to feed themselves on waste lying on the road.

Some attempts have been made at a local level to regulate illegal dairies and mandate proper care of cattle. City authorities are usually directed to impound street cows, release them to owners for a fee, and auction or rehome unclaimed animals. Some local governments have their own cowsheds, where confiscated or ownerless cattle are taken to live out their days. These "retirement" sheds are not necessarily a happy ending either. Twenty years ago, the Delhi government created 10 such shelters. Now only five survived, amid complaints of a lack of funding or oversight. Out of 49,000 cows kept in these shelters, authorities reported in 2011, 46,000 died in three years. Many of the animals are severely malnourished by the time they reach a new home. Trash, plastic, shoes and even metal is found in their stomachs.

Problems of Impounding Stray Cattle

The problems of impounding stray cattle are many: 1) The paucity of trained man power with the Municipal Corporation to catch and load them in a truck. 2) Lack of Gaushalas with adequate space. 3) The tendency of the neighbouring towns/cities to catch their stray cattle and leave them outside near the next city.

Stray Cattle Menace in Delhi

Life is endangered by stray cattle menace in Delhi. Elderly, women and children are often victim of stray cattle roaming on the roads. Drives launched by MCDs against stray cattle and un-authorised dairies across the capital have called off under pressure from the infuential dairy owners' lobby and politicians. Even Delhi High Court directives have failed.

Stray animals including cows, buffaloes and bulls can be seen roaming on the city roads and the most affected areas are those located near urbanised villages, illegal dairies, slums, unauthorised colonies and parts of Outer Delhi. Due to lack of cattle-rearing fields, cattle owners leave their animals to roam on the streets.

While the cow menace on roads and a berserk bull have made news, Kotla Mubarakpur in Delhi tells a story of neglect and sickness. Often cows which are too old to be of any commercial value to their owners, are simply abandoned. Every now and then, a dead animal can be spotted alongside the drain. Residents of the area let cows out to graze. Rather than any grass, it is the garbage from the Naala that the animals munch on. The resultant effect of this on the milk that they give is best left to one's imagination. In winters, things get worse. There is no shed to shelter the cows, resulting in their death due to the cold weather.

No Space for Cows at Delhi's 5 Shelters

For want of space, one can witness stray cattle blocking roads in different parts of the city, and in many cases even causing accidents. The problem could worsen further as five major cow shelters, aided by the Delhi government, deny admitting any more cattle due to space crunch.

The three corporations of Delhi- North, South and East- catch stray cattle and send it to one of the shelters. The Ghuman Hera Gramin Gaushala in Najafgarh has maximum space for 2500 cows, Sri Dabur Hare Krishna Gaushala in Sureda village can accommodate 1500-2000 cows. In 2002, the Supreme Court ordered authorities to rid Delhi's streets of cattle and shut all dairies functioning out of urban areas so that abandoned animals do not find their way to Delhi's busy streets (Source : TOI-22.06.2017).

Initiatives taken by Delhi Government to Curb Stray Cattle Menace

Delhi government hired about 150 stray cow catchers who are supposed to bring 10 cows per day. Due to the corruption, these stray cattle often make their way back on street and are often caught multiple times. The other solution lies in relocating the dairies, centrally, outside the city. The authorities have to create infrastructure for these dairies so that the owners can operate them. Creating huge Gaushalas, at appropriate locations, can take care of stray cattles.

Steps to Curb Stray Cattle Menace in Delhi

The High court has been directing civic agencies to free Delhi from Stray cattle for long. On May 26, 2005 the court had given four weeks to MCD to free east Delhi from stray cattle. The Bench, which is monitoring the progress in this regard on weekly basis, had earlier taken strong exception to non-implementation of its order passed in 2003 wherein several directions were issued to the MCD.

Keeping in view the gravity of stray cattle menace, High Court on 13 July, 2005 reminded the Delhi government, MCD and NDMC to solve the cattle menace and directed the agencies to file a joint status report within a week. If the agencies fail, their senior most officials will have to appear in the court on July 21. Another deadline by the Delhi High Court to rid the city of the cattle menace by August 30, 2008 could not meet by civic agencies. Even spending huge amount of money by veterinary department on relocating stray cattle did not yield any positive results. We still see so many cattle on the roads. Defending itself, the veterinary department blamed the dairies, which are operating illegally in the city, for the menace. According to an estimate by the civic body, there are close to 2,361 illegal dairies running in 2011 (Source: HT 24.03.2011) in the city. A majority of the dairies, which have been allotted a site at the Ghogha Dairy, following a court order in 2002, are yet to be shifted. The only solution is to relocate them to the allocated site.

Cattle Menace in Parched Bundelkhand

Almost 80% of Bundelkhand's 1.83 crore population lives in villages (2011 census) and much of this depends on farming. Livestock, reared as insurance against crop failure, is traditionally left for grazing during the lean season. Water shortage during summer (June 2016) had killed nearly 3 lakh cattle. Farmers who don't have enough to feed themselves, how can they afford to keep animals.

Why Farmers Abandon Old and Unproductive Cattle?

The reasons are obvious. Farmers can neither sell nor maintain their unproductive animals. If an animal dies, farmers have to pay ₹2,500 to ₹3,000 to bury the carcass. People somehow get water to drink, but getting water for cattle is tougher. The stray cattle problem has increased the load on Gaushalas which now struggle to tackle the increased load. The ban on sale has put farmers in a fix. Unless the old pair is sold, the farmer doesn't have money for a new one. Maintenance of unproductive cattle is a bigger issue.

Stray cattle are not a choice but compulsion. Governments of most states have banned cow slaughter and beef. And the Centre too prohibits the sale of cattle for slaughter in animal markets. It's difficult for a farmer to feed unproductive livestock. Besides, stricter norms for transporting cows have left people with the little option but to abandon them.

Why Are Indian Cities Infested With Stray Cows But Not Stray Buffaloes?

There are reasons behind it. 1) Buffaloes in India are more expensive than cows. So they are more valuable commodity for dairy farms. 2) The major reason why cows land up on streets but buffaloes don't have to do with Indian culture. Once cattle stop giving milk it is of no use for dairy farms. In case of buffaloes they sell it to slaughter houses for meat and skin but in case of cows there is a religious fear. In many cases they just leave them on streets to fend for themselves instead of risking committing a sin by being part of cow slaughtering. 3) The cows have the ability to remember the way home unlike buffalo and other animals. It can track the way home easily so it is a blessing to owner as he will have the cattle ready at the door step for milking in the evening. Again buffalo is a different animal. It will never reach its home in evening. 4) Moreover hindu religion treats cow as a sacred animal so chances of somebody hurting a cow is very less. All this factors combined makes it an ideal animal to roam on the Indian roads.

Cows should be taken care of in a responsible manners. India's roads should be free of cows. Animal owners should take proper care of their charges and those animals that are stray, should be taken care of properly - neutered if necessary. India needs to improve its infrastructure if it's to maintain its march towards a brighter future. This cannot be done if stray animals, sacred or otherwise, are clogging up its roads. Let us return the cow to its rightful place, culturally, religiously and economically. Let us not misunderstand the past for the present and let not the sacred become something we, as Hindus, as scared to talk about.

5

Vigilantism and Violence

Cow vigilante violence involving mob attacks in the name of "cow protection", has swelled since 2014. Cattle slaughter is banned in most states of India. Recently emerged cow vigilante groups, claiming to be protecting cattle, have accused some Indian Muslims and Dalits of cattle theft or slaughter, and targeted violence against them, leading to a number of deaths.

Cow vigilantism itself is not new in India, and violence over the protection of cows has occurred in the past. However, the frequency, impunity, and fragrance of the current instances of cow-related violence are unprecedented. In 2015 Business Insider reported that vigilante attacks on trucks carrying cattle had increased in Maharastra. In 2017, Bloomberg reported that according to the meat industry representatives, cow vigilantes have been stopping vehicles, extorting money and stealing valuable livestock.

As of 2016, cow protection vigilante groups were estimated to have sprung up in hundreds, perhaps thousands of towns and villages in northern India. One kind of cow protection groups are gangs who patrol highways and roads at night, looking for trucks that might be smuggling cows across the state borders. The gangs consist of volunteers, many of whom are poor labourers. The volunteers often tend to be young. The vigilantes often have a network of informers consisting of cobblers, rickshaw drivers, vegetable vendors etc., who alert them to supposed incidents of cow slaughter. The group members and their network often use social media to circulate information. Some vigilantes claim to work with the police.

Laws and Legal Issues

The government has placed a number of restrictions on the slaughter of cattle. In May 2017 government banned the slaughter of cattle for purpose of exporting beef. This restriction threatened an Indian beef export industry worth $4 billion annually. Several Indian states have tightened restrictions on the slaughter of cows. For example, in March 2015, Maharashtra passed stricter legislation banning the sale, possession, and consumption of beef. Some Indian states have been accused of having laws that enable cow protection groups. In April 2017 the Supreme Court asked the governments of six states: Rajasthan, Maharashtra, Gujarat, Jharkhand, Karnataka and Uttar Pradesh; to respond to a plea asking for a ban against cow-protection related vigilantism, due to the violence perpetrated by these groups.

A number of incidents of violence have occurred since 2014. According to a June 2017 Reuters report, citing a data journalism website, a total of 28 Indians - 24 of them Muslims, have been killed and 124 injured since 2010 in cow-related violence. The frequency and severity of cow-related violence have been described as unprecedented. Incidents of violence in chronological order are given below.

5 April 2017, Haryana: In the Alwar mob lynching, Pehlu Khan, a dairy farmer from Nuh district of Haryana, was murdered by a group of 200 cow vigilantes in Alwar, Rajasthan, India. Six others who were with Pehlu Khan were also beaten by the cow vigilantes.

24 April 2017, Jammu and Kashmir: A family of five, including a 9-year old girl, were attacked and injured.

20 April 2017, Assam: Two men, in their 20s, were allegedly killed by a mob of cow vigilantes, after being accused of trying to steal cows for slaughter.

23 April 2017, Delhi: Three men transporting buffaloes were injured after being beaten by a vigilante group.

23 June 2017, Delhi-Ballabhgarh train: Four Muslims were lynched allegedly over rumors of eating beef.

1 May 2017, Assam: Two Muslim men were lynched in Nagaon district of Assam on suspicion of stealing cows.

22 June 2017, West Bengal: Three Muslim men were lynched in Islampur, Uttar Dinajpur for allegedly trying to steal cows.

27 June 2017, Jharkhand: Usman Ansari, a 55-year-old Muslim dairy owner, was beaten up and his house set on fire by a mob in Giridih district.

29 June 2017, Jharkhand: Alimuddin, alias Asgar Ansari, was beaten to death by a mob in the village of Bajartand, allegedly for carrying beef.

10 November 2017, Alwar, Rajasthan: 2 Cattle traders named Ummar Khan and Tahir Khan were allegedly thrashed and fired at by cow vigilantes. Ummar Khan died on the spot due to bullet wound and Tahir Khan was admitted to a hospital.

11 July 2016, Gujarat: A group of six men attacked four Dalit men after finding them skinning a dead cow. The four were chained to a car, stripped, and beaten with iron rods.

30 July 2016, UP: A mob attacked the house of a Muslim family on suspicion of cow slaughter in UP's Muzaffarnagar district.

26 July 2016, MP: Two Muslim women were beaten up at Mandsaur railway station in Madhya Pradesh on suspicion of carrying beef.

11 July 2016, Gujarat: Seven members of a Dalit family were beaten up by about 35 gau rakshaks for skinning a dead cow in Una in Gujarat's Gir Somnath district.

10 July 2016, Karnataka: Bajrang Dal members brutally attacked a Dalit family in Koppa in Karnataka, claiming there was beef inside their house.

10 June 2016, Haryana: Two beef transporters were beaten up and force-fed cow dung allegedly by the Gurgaon Gau Raksha Dal in Haryana.

6 May 2016, Maharastra: The Bombay High Court ruled that it will no longer be illegal to consumer or keep imported beef but upheld the Maharashtra government's ban on slaughter of cows and bullocks.

2 April 2016, Haryana: Mustain Abbas was killed, allegedly by Gau Raksha Dal members in Kurukshetra, Haryana while he was transporting a buffalo. The High Court of Punjab and Haryana ordered CBI to probe the murder on May 9.

18 March 2016, Jharkhand: Two muslim cattle-traders including a 15 year old were beaten and hanged from a tree in Jharkhand's latehar district. Five suspects, including a member of a local gau raksha vigilante group, were later arrested.

13 January 2016, MP: Cow protection group attacked a couple at Khirkiya railway station in Madhya Pradesh over allegations that they were carrying beef.

3 December 2015, Haryana: Violent clashes broke out in Haryana's Palwal district when villagers stopped a truck allegedly carrying cow meat.

19 October 2015, Delhi: Hindu hardliners threw black ink at J&K MLA engineer Rashid in Delhi. He was thrashed earlier for holding a beef party in Srinagar.

16 October 2015, Himachal Pradesh: A mob laynched a man to death in Sirmour district of Himachal Pradesh for alleged cattle smuggling.

14 October 2015, Himachal Pradesh: A mob beat a 22-year old to death, and injured four others, after suspecting them of transporting cows.

9 October 2015, Jammu and Kashmir: A right-wing Hindu mob in Udhampur district threw gasoline bombs at an 18-year-old trucker. The mob had incorrectly suspected the trucker of transporting beef.

9 October 2015, Jammu and Kashmir: A petrol bomb attack on a Srinagar-bound truck in Udhampur left three men- two kashmiris and a policeman - battling burn injuries. The truck was set on fire allegedly by rightwing activists.

9 October 2015, UP: A mob went on a rampage following rumours that a cow had been slaughtered in Mainpuri district in UP.

6 October 2015, Karnataka: A cattle trader in Karnataka had a narrow escape after Bajrang Dal activists attacked him with metal rods on a rumour about a stolen cow.

1 October 2015, Kerala: Six students of Sree Kerala Varma College in Thrissur, Kerala, were suspended for organizing a beef fest on campus to protest against the Dadri lynching.

28 September 2015, Uttar Pradesh: In the Dadri lynching, a mob of villagers attacked the home of a Muslim man Mohammed Ikhlaq, with sticks and bricks, who they suspected of stealing and slaughtering a stolen cow calf, in Bisara village near Dadri, Uttar Pradesh. 52-year-old Mohammad Akhlaq Saifi (Ikhlaq according to some sources) died in that attack and his son, 22-year-old Danish was seriously injured.

29 August 2015, Delhi : Residents of Chilla village, near east Delhi's Mayur Vihar, clashed with four truck drivers who were reportedly transferring buffaloes to a slaughter house in Gazipur.

2 August 2015, Uttar Pradesh: According to Human Rights Watch, "purported animal rights activists allegedly belonging to People for Animals" beat three men to death, after the victims were found carrying buffaloes.

30 May 2015, Rajastan: A 60-year old man who ran a meat shop was beaten to death by a mob with sticks and iron rods.

16 March 2015, Maharashtra: Haryana passed a stringent bill banning the sale of beef. The law mandated five years of rigorous imprisonment for selling beef - same as that in Maharashtra- and a fine of up to ₹50,000.

4 March 2015, Maharashtra: Maharashtra banned beef; anyone found in possession or sale would face five years of jail and a ₹10,000 fine. While a 1976 law prevented the slaughter of cows in Maharashtra, the new Act banned the slaughter of bulls as well as bullocks.

Vigilantes and Violence in Rajasthan

According to police, in 2017, a total of 389 cases were registered under the Rajasthan Bovine Animal (Prohibition of Slaughter and Regulation of Temporary Migration or Export) Act, 1995. Of these, 259 were charge sheeted while 24 cases were closed and 76 still pending. In 2015, 543 cases were registered while in 2016 the number dropped to 474. The cases registered under Bovine Act in Rajasthan in 2017 are mentioned below:

Cases Registered Under Bovine Act in Rajasthan (2017)

District	No. of Cases
Bhilwara	1
Tonk	3
Alwar	3
Sikar	1
Jhunjhunu	1
Dholpur	1
Barmer	1
Udaipur	1
Pratapgarh	1
Chittorgarh	3
Shri Gangangar	5

(Source: HT 08.12.2017)

On November 10, 2017, the mutilated body of 35 year old Umar Mohammad of Ghatmika was found on the railway tracks. Police in Rajasthan's Alwar also shot at a truck being allegedly used to smuggle cows on 11 December 2017. This is the second such case, a man was killed when police in Alwar similarly opened fire to stop a truck. A third incident took place on 25 February 2018 where a body of a man, suspected to be a "cow smugglar" was recovered from an abondoned vehicle having 23 cows, five of which were dead.

Cow Vigilantes and Violence in Maharashtra: Three youths were allegedly thrashed on 26 May, 2017 in Washim district of Maharashtra on suscipicion of possessing beef by cow vigilantes (Source: HT-30.05.2017).

Cow Vigilantes and Violence in IIT Madras: An IIT Madras scholar who was participating in a campus beef festival was assaulted by a group of students in June 2017 (Source: HT 01.06.2017).

In most cases, the issue of cows has been the proximate cause - allegations of cow slaughter, smuggling, eating or even possessing beef. The spread of these cow terrorism cases - 12 states in all is chilling between June 2014 and December 2015, 11 such attacks took place, but after that, the pace has increased with 2016 recording 12 cases and 2017, 9 cases in six months. Most such attacks have occurred in North India.

Impact of Vigilantism

India has been beset by a wave of gruesome lynching and the epicenter of the country's violent upheaval is the indolent cow. In the last one year, ***Gau Raksha*** Dal has become a state gifted passport to tens of thousands of lumpen goons to vent their sadism upon the poorest and most helpless people of this country. In towns and villages across the country groups of young men have begun to roam the streets and country roads looking for people transporting meat, live cattle, or the hides of cattle that have died of starvation. An estimated 200 such groups are believed to be active in Delhi alone. Another 200 or more have sprung up in Gujarat. Their usual mode of operation is to stop trucks carrying livestock or hides, accuse the owners of having slaughtered cattle, beat them up and release them after they hand over all the cash they possess. Criminals have turned cow protection into a business, and have given this noble task a bad name.

The atrocities have steadily been mounting. In September 2015, Mohammad Akhlaq was hanged over rumors that he killed a cow and refrigerated its meat. A month later, 16-year-old Zahid Rasool Bhaat was slain by vigilante groups. In March of this year, suspected cattle traders Muhammed Majloom and Azad Khan were lynched. In April, 55-year old dairy farmer Pehlu Khan was accused of smuggling cows and was brutally beaten to death. In May, traders were assaulted for alleged beef storage, and Abu Hanifa and Riazuddin Ali were killed for purportedly stealing cattle. In June, Ainul Ansari was attacked on suspicion of transporting beef, while 15-year-old Junaid Khan was stabbed to death by a mob after being branded a beef eater. Since September 2015, there have been more than a dozen lynchings across the country.

Vigilantism and Violence in the Name of Religion: The people are getting lynched or hacked to death over cows and at times women. As if these are interchangeable commodities in a patriarchal marketplace. Murder has replaced dialogues. The latest video from Rajasthan is unbearable, a muslim man crying out for mercy- babu, babu, babu, please don't kill me - while being hit repeatedly with a pickaxe and then set on fire (Source : Outlook : 18 Dec. 2017). Will this madness ever stop? The dusty loose soil of the Ganga-Jamuna plains seems to have an insatiable thirst for blood. How much longer will the Muslim, Christian and Hindu rulers take turns in subjugating the other? Has our world become a asylum or have we gone mad in this new world? Hatred for votes, hatred for money and hatred for fame! Everybody loves to hate someone. The Bhakts hate liberal elites, the liberal elites hate bhakts and both hate people trying to find some middle space. Yes, there is no middle space in a murder scene nor is there any logic to religion.

The recent horrific lynching of a young man, Junaid Khan, in a train near Ballabgarh on the out-skirts of Delhi is another instance of a growing trend of mob violence against members of the muslim community in India. The target of cow vigilantes has extended to Dalits and tribals too and in some cases to police personnel and government officials. One such incident that shocked the country took place a year ago in Gujarat where four dalit youths were stripped and publicly flogged even as a video was shot of the incident.

Cow Vigilantes and Violence in Karnataka: Tension prevailed in the communally sensitve port town of Honnavar on 8 March, 2018 after about 150 self proclaimed 'Gau Rakshaks' chased and thrashed two men for transporting cattle allegedly to a slaughter house in Bhatkal on 7 March, 2018 night. These cattle were legally transported from Ankola to Bhatkal. In December, Honnavar also witnessed riots after a youth of 18 years age was found dead. He allegedly died while trying to escape a stone-pelting mob in Honnavar. Several police vehicles and public property were destroyed in clashes.

Why Politics on Beef

India is known for diversity and tolerance. So long India maintains its unique feature of diversity and tolerance, its contribution to the world and the society will be much greater than just the political. We should have a secular and progressive approach.

Almost 95% beef traders are Hindus, still a man was lynched in Dadri because he ate beef. This is the death of mankind and humanity. Eating habits have nothing to do with religion. MPs and MLAs, too own beef companies. Then why is only the common man being targeted. Banning cow slaughter is not the way to create a new arcadia for the beleaguered Indian cow. In an atmosphere vitiated by communal politics, the cow has become a political animal.

Cow Vigilantism Leads to Economic Terrorism

Cow protectors (Gau Raksha Dal) used to terrify anyone dealing with the holy cow. Economic terrorism has been introduced by targeting of anyone who deals with cattle, cattle skin or animal byproducts. GRDs are out to deprive some of India's largest (and poorest) communities such as Dalits and Muslims who are worst affected. There have been incidents all over north India in recent months. The vigilantism is spreading. In Punjab and Haryana, GRDs have brought prosperous elite- caste Hindu merchants to their knees by levelling false charges of trading in cow by-products. GRDs function like organised extortion rackets. Even hides of fallen cows headed for tanneries in UP are unsafe. GRDs, aided by police, routinely extort wayfarers in Gaumata's name. What's forgotten is that 60 percent of the value of cattle is for other uses, such as: leather for domestic and industrial use, glue from bones, collagen and elastin for the cosmetic industry and so on. GRDs are also attacking buffalo, goat and even chicken transporters and porcessors, as they aim to impose vegetarianism and capture the cattle trade. The GRDs are wrecking a whole network of downstream industries, like tallow manufacturers who provide raw material for soap. The chain of production that yields cattle products provides subsistence to millions of livestock producers (farmers, dairies), traders, butchers, wholesale meat dealers, retailers and dealers in industrial products.

Vigilantism - A Threat to Cattle Economy

Across the country, especially in rural areas, there's simmering unease bordering on panic. Farmers no longer think freely about buying or selling cattle. People, especially from the minority communities, can't easily think about meat of any kind. Truckers are worried about transporting cattle including buffaloes. Dalit leather artisans live in terror. All this because of small vigilante groups taking upon themselves to prevent "cow slaughter" through violence.

This isn't just infringement of individual rights. The so-called "cow protection" vigilantism is taking a toll on the leather and meat processing industries, destroying livelihoods of nearly 8 lakh Dalit leather artisans, among the most oppressed sections of society, and also snatching valuable nutrition from vulnerable tribal communities of the Northeast. It's also hurting farmers as they can't sell their cattle. In fact, they're finding it difficult to dispose of corpses of fallen (naturally dying) cattle. All this is pushing up the numbers of stray cattle, which stand at 5.3 million. They face a slow death due to starvation and, increasingly, from eating plastic in urban waste dumps.

There are over 8 lakh flayers spread across the country who flay hide from carcass, preserve it and sell it to middlemen. Flayers come from traditional leather working castes, extremely poor and socially stigmatised. The periodic work of flaying brings in much-needed income. Because of sporadic violence, often targeting leather workers, flayers now face a severe shortage of work, besides the threat to their lives. At the other end of the value chain, the leather industry, a big foreign exchange earner, is in crisis. Leather goods export will be affected as supply for raw hides falters.

Impact of Vigilantism on Beef (buffalo) Export

Official data shows that while 1,37,687 tones of beef was exported in March 2017, the figure came down to 86,119 tones in April 2017 - a drop of 51,568 tones in just a month. The recent attacks by vigilantes on cattle transporters coupled with the crackdown on illegal slaughter-houses, has hit the industry badly. The trade has gone down by almost 50%. The latest notification of the Union government banning the sale of cattle for slaughter has added to their woes. The business has gone down by ₹600 crore, which is nearly half of their business. Virtually movement of meat from slaughter house has ground to a halt across the states for fear of vigilantes (Source: HT - 07.06.2017).

Importers from Germany, France, Italy and the US, unsure about supplies from India, are looking to other players such as Pakistan. The industry is experiencing a 15% slowdown because of a shortage of hides. Though cow slaughter is legal in Bengal, which accounts for nearly half of leather exports, it sources hides from states like UP. Bengal generates only 35% of the hide processed here. Units with export commitments are in a fix. As deliveries stall, their goodwill gets diluted. Exports to Germany have fallen by 24% and France 20%.

Vigilantism Threatening Law and Order

Cow vigilante groups are now a law unto themselves, assaulting and killing alleged cattle smugglers, gheraoing police stations. Since September 2015, when a Muslim was lynched near Dadri on Delhi's outskirts for allegedly possessing cow meat, vigilantism has become rampant particularly in states such as Jharkhand, Punjab, Haryana, Rajasthan and UP. The shocking incident in Jharkhand's Latehar district, where two alleged cattle traders were thrashed and hanged, made front page news in March, 2016. But many cow vigilantism incidents go unnoticed. In the last six months, harassment of cattle traders has increasd manifold.They're beaten up on the name of cow protection. Fear has made many abandon their trade and are jobless (Source: TOI-26.06.2016).

Court Convicts 11 Cow Vigilantes in Jharkhand's Lynching: A fast track court in Jharkhand's Ramgarh district awarded 11 people life imprisonment on March 21, 2018 for the killing of 55-year-old Ansari, a landmark verdict because it was the first in a case connected to a violence by vigilante groups in the name of 'Gao Raksha'. Ansari was killed on June 29, 2017 **(Source: TOI - 23.03.2018).**

Protest against Vigilantism: After an attack on four Dalits in Gujarat in July 2016, thousands of members of the Dalit community took to the streets to protest. The protests spread across the state. In clashes with the police, one policeman was killed and dozens of protesters were arrested. At least five Dalit youths attempted suicide, one of whom died.
A campaign was conducted through a Facebook post against the violence. A lot of people took out a morcha at Jantar Mantar in Delhi and 16 more cities across the country, including Mumbai against mob lynching in the name of cow vigilantism. International organisation, Human Rights Watch in April 2017 reported that Indian authorities should promptly investigate and take action against the self-appointed "cow protectors", who have carried out attacks against Muslims and Dalits over rumors of selling, buying or killing of cows for beef.

Cow Vigilantism is not justified: Cow vigilantism in India has become lawlessness under the rubric of cow protection. The term "vigilante" refers to a self-appointed person or a group of persons that undertakes to enforce the law without any legal authority. It includes persons who take the law into their own hands to avenge what they may perceive to be crime. Murder and violence of men by other men is illegally and extra-judicially enforcing religious beliefs. Cow vigilantism as an expression has sadly become a convenient cloak for violence, brutality, murder and mayhem in the name of the cow. The truth is that in this immense profaning of the sacred underway in the violence in India, India is rendered a human abattoir - a slaughterhouse of men and women to save cows from slaughter.

Cows are sacred to many; and one respects that. But violence by humans upon humans - regardless of the reason being sacred or profane is inhumane, in any civilised society. That human life is sacred; that the holiness of the cow mustn't supersede the sanctity of human life. Once again, the slaughter of men in the name of cows threatens to slash and slice the soul of India. In the matters Sacred or Profane, India can no longer afford to be a mute spectator to the brutal assassination of its multi-religious, multi-ethnic and multi-racial soul. It must speak out, speak up and, for all to see, stand up for its soul. The people cannot be allowed to take law in their own hands in the name of religion because the consequences can be disastrous for the social fabric of the country.

Need for Urgent Steps to Stop Cow Vigilantism

Dalits and Muslims have been at the receiving end of a rash of violence unleashed by lynch mobs, especially in the northern States. Non-violence is the founding faith of this country. The Centre cannot turn its back on the violence. The States have the responsibility to lodge FIRs against these vigilantes. As most of the violent incidents had occurred on highways, the court directed the States should take steps to initiate highway patrolling. The court also directed the Chief Secretaries and the Directors General of Police of States to consult each other and respond to the court. The Centre shall also indicate its views on this issue. The government have denied supporting cow slaughter vigilantism. In August 2016, PM also condemned it.

The Supreme Court directed the Centre and the States on Wednesday to take stern action to curb cow vigilantes by ensuring they don't terrorise highways and sought the appointment of nodal officers in each district to coordinate measures against self-styled "Gau Rakshaks". The cow vigilantes should not be permitted to take the law in their hands or become a law unto themselves. Some mechanism has to be there to prevent violence by these groups. This must stop. Some kind of planned and well-coordinated action is required by the governments so that vigilantism does not grow. The court directed state governments to appoint a senior police officer for each district as the nodal officer to coordinate action against cow vigilante groups. If any kind of deviancy takes place, the said nodal officer shall take action and such vigilantes will be booked in accordance with law, the court ordered.

Why Violence in the Name of Cow Protection?

Economics, not religion drives ownership of cattle in India. The increasing incidents of violence under the garb of cow protection in the country are driven largely by the belief that Muslims engage with the cattle economy mostly for meat (as butchers, commission agents or beef eaters). Viewing the cattle economy in the country through the prism of religion, makes no sense. Non-Muslims also eat beef in India, although Muslims account for the majority of beef and buffalo meat eaters in the country. Around 63.4 million Muslims (or 40% of the Muslim population in the country) eat beef or buffalo meat. Overall, 80 million Indians eat beef or buffalo meat. It emerges that economics rather than religion drives cattle ownership in India. The cattle ownership shows a similar pattern across religions. The National Sample Survey Office (NSSO) conducted the All India Debt and Investment Survey (AIDIS) from January to December 2013. The survey looked at more than 110,000 households in both rural and urban India and collected information on cattle ownership by households in different states.

In Kerala, 7.5% and 5.5% of Hindu and Muslim households respectively own cattle. These numbers are 52% and 21% respectively in Uttar Pradesh. In Jammu and Kashmir, 57% Muslims own cattle. For Hindus in Jammu and

Kashmir, this number is 37%. State-specific economic and climatic conditions have a role to play in cattle-owning decisions. It shows that livestock decisions are based on economics rather than a love for cows. Only 15% of households own non-milk animals in India. This is half of the overall cattle ownership figure of 30%. Again, the figures are not very different across religions. Where the non-milk bovines end up can be guessed by anybody. They are the suppliers for India's multi-billion dollar beef industry. It has been an unacknowledged but convenient arrangement. Most owners sell their non-milk cattle without asking questions about the end use. This is why the local cattle trade is crucial for India's livestock economy. A government ban on sale of animals for slaughter in local markets, and vigilante mobs attacking those transporting cattle in the name of gau raksha, can destroy this arrangement.

Impact of Cow Vigilantism

How vigilantism is killing the country's largest economy of the poor? Forget about the loss of revenue from exporting buffalo meat after government's ban on procuring animals from market place for slaughter, the real impact will be on India's poorest farmers, and the economy of livestock which is worth more than ₹3 lakh crore. In just a decade, there is a historic shift in the agriculture sector. Economic contribution of livestock is today more than that of food grain crops. Traditionally, of the three components of the sector—crops, livestock and fisheries—crops drove the growth, and food grains are a major part of it. As a result, policy and programmes focused on crops. It has been more than a decade that livestock economy is booming as a result of farmers' conscious decision to switch over from crops. A report prepared for the 12th Five Year Plan (ended in 2017) accepted this shift by recognising livestock as the engine of agriculture growth.

Livestock controls close to a quarter of the agriculture gross domestic product (GDP). In 2010-11, it generated outputs worth ₹3, 40,500 crore (at current prices). This was 28 per cent of the agriculture GDP and about 5 per cent of the country's GDP. Rise of the livestock sector has implications for poverty. Rural poverty is less in states where livestock contributes more to farm income. Punjab, Haryana, Jammu & Kashmir, Himachal Pradesh, Kerala, Gujarat and Rajasthan are a case in point.

With all these regulations and confusion over government's own policy towards cattle, the new economy of the poor is under grave threat. As unemployment grips India, the livestock sector could have been a natural choice to tide over the crisis. But if one farmer is not sure of selling a cattle, and not sure about his or her security while dealing with cattle, the economy will just collapse. Like in the free market, sentiment also is critical for the economy of the poor.

Farmers find it hard to keep them, traders aren't interested in buying them, and gaushalas are full. Once a valued resource, cows are now becoming a farmer's worst enemy in many places. As a deterrent, some farmers have installed electrified barbed wire fencing, even though it is illegal. Even though the problem of stray cattle is not new in India, in the last few years, it has reached highly alarming proportions. According to 2012 data from the 19th livestock census, stray cattle amounted to 58.87 lakh and this number may go up dramatically in the 20th livestock census which will be out later this year. Since the Centre notified the Prevention of Cruelty to Animals Rules in May, it has become even more difficult to buy, sell or transport cattle. Cow vigilante groups have dramatically changed the livestock business. The practice of unproductive cattle winding up in slaughterhouses and being used for meat and leather was already in trouble. Until recently, farmers followed a pattern where they sold their old, unproductive cattle to traders, who then transported them to states where slaughter is allowed. The farmers, on their part, would buy young and productive bovines from cattle fairs. But this has become a habit of the past now. Farmers simply abandon their unproductive cattle, because maintaining them is unaffordable.

Old Cattle are abandoned and dying of malnutrition: The attacks by cow vigilantes on transporters ferrying cattle, coupled with tough rules on cow transportation, have forced most farmers to abandon the idea of buying cattle. The gaushalas, meanwhile, are full beyond capacity. Hingonia gaushala, on the outskirts of Jaipur, is struggling to look after more than 14,000 cows, which is almost twice as much as it can accommodate. And yet, municipal authorities keep bringing in more animals. Last year, over 8,000 cows in the gaushala died of malnutrition and disease.

6

Cow Shelters are Becoming Cow Killers

There are about 4,000 Gaushalas in the country, of which 1,325 are run by various animal husbandry departments of the states. Majority of the Gaushalas in the country are poorly run where the old and abandoned animals are herded in one small enclosure. The average mortality rate in Gaushalas is 10% a month. The aim should be to make shelters more sensitive to the need of the animals that are sought to be saved and also make them financially viable.

Cow is a holy animal for the Hindus and over the past three years, India has seen an increase in attacks by self-styled cow vigilantes. Their aim is to stop cow smugglers and those who illegally trade in beef (cow slaughter is illegal in several states), but the victims have sometimes been innocent cattle traders and those who skin the carcasses of dead animals.

With farmers finding it unviable to look after their unproductive and aged cows, many of these animals land up in Gaushalas. Though, many of these Gaushalas are in poor condition, there has been no effort to prescribe standards for them. Gaushala owners manhandle cows in the shelter. After sometime Gaushalas degenerate into semi-dairies. The group controlling the Gaushalas starts segregating the milking cows from the old and sick ones. These are fed better, made to breed and the milk is collected and distributed among these people. The calves are often sold to butchers. The living condition in majority of the shelters, even those run by the government agencies are horrifying. The animals are not fed properly, live in unhygienic condition and most do not have qualified veterinarians.

In other countries when animals grow old they are exterminated. Here we do not kill them but keep them in appallingly unhygienic condition. After a long cruelty, animals finally die. Such gaps in respect of how to construct an animal shelter, the kind of food to be served, from protocols to rescue to disposal of carcasses, from the kind of veterinary services that should be offered to the manufacturer of eco-friendly products from cow dung needs to be addressed. Creation of modern sheds, biogas plants and solar units also need to be addressed. The condition of cow shelters in most of the states in the country is pathetic. Cow Shelters in few states are cited below.

Lucknow Kanha Upvan Gaushala

499 stray animals-most of them cows have died in Lucknow's Kanha Upvan from April 1 to September 30 in 2017. Fund crunch was the main cause of deaths. Though ₹8.35 crore aid from the state government in the last five years (2012 to 2017) was given to the cow shelter. 86% of the total grant given by the Uttar Pradesh Gau Seva Aayog to cow shelters and cattle protection organisations went to an NGO Jeev Ashray, which runs Kanha Upvan Gaushala near Amausi, owned by Lucknow Nagar Nigam. Over five years, from 2012 to 2017, the Aayog distributed grants worth ₹9.66 crore, of which ₹8.35 crore went to the NGO alone (Source: TOI-03.07.2017).

Kanpur Gaushala

152 cows died of starvation in UP's richest cow shelter. Four of these deaths at the 128 year-old Kanpur Gaushala occurred in last week of March 2017. As per doctors, statement, cows starved to death, triggered furore among local residents. The shelter holds 540 cows, 152 of whom have died. Four of these deaths occurred in the first week of April 2017. The society that manages the facility holds property worth more than ₹220 crore. The Kanpur society was formed with the sole criterion of taking care of stray cows but allegations of neglect and siphoning of funds were triggered after an autopsy of the four dead cows pointed at empty intestines and urinary bladder. It means the cows hadn't eaten anything nor did they drink water out of weakness (Source: HT-08.04.2017).

Recently, the Uttar Pradesh Government decided to set up cow shelters in prisons with 12 jails (Source: HT-07.02.2018). The prisons will be entrusted to take care of the cattle from feeding, cleaning to milking in addition to their assigned work in the premises. Within a week, the Government also announced to construct Gaushalas in all 75 districts of U.P. The Goverment decided to set up a cow protection commitee in each district under the chairmanship of the district magistrate concerned (Source: HT-11.02.2018). It is easy to announce anything by the politicians, may be to bring heaven on earth, but we must not forget shocking state of affairs in two big Gaushalas of Uttar Pradesh (located in Kanpur and Lucknow). Hundreds of cows died within few months in these Gaushalas in 2017.

Cow shelters in Gujarat

Most of the shelters and their inhabitants in Gujrat are starved of money and with successive years of drought leading to cattle being abandoned in large numbers, space at the shelters is also short. Each Taluka has two cow shelters on average. Some 6,60,000 cows are estimated to be in these shelters across the state and struggling to survive without government support.

A large number of cattle is more or less left to die at these shelters, at a time when cow vigilantes are virtually running amok with the goal of saving the animal - considered sacred in the Hindu religion. The government is more interested in helping cow protectors than cows. It's because cow protectors give the government political mileage. They will not let Dalits skin dead cattle because it hurts their sentiments. They will not let Muslims trade the animals because the cow is their mother. But they have no problem when these animals die of starvation in the pinjrapole (Source: HT 23.07.2016).

Hingonia Cow Rehabilitation Centre in Rajasthan

500 cows died at a Rajasthan government run cattle shelter near the state capital within two weeks in August, 2014. The deaths took place at the Jaipur Municipal Corporation's Hingonia Cow Rehabilitation Centre. All the 266 contractual workers at the establishment were on strike since July 21, 2014 because they hadn't been paid their dues for May and June. The bovines starved to death after they were caught neck-deep in cow dung and fodder that had turned slushy due to rainwater seepage. Bad planning turned the place into a quicksand of sorts. The cows died after being trapped for days without food and water (Source: HT 06.08.2014).

Delhi's Gaushalas

Delhi has no time for its cows. Illegal dairies in urban areas abandon animals that are useless to them, even milch cows left to fend for themselves through the day. Cattle are dying by the thousands in Delhi's cow shelters, even as protection groups and political parties squabble over saving the 'holy' cow. Data collated by the municipal corporations in Delhi and government cow shelters is startling. The South Delhi Municipal Corporation sent 3,398 cattle to shelters in 2015-16 and 3,685 died during the same period. In 2014-15, 2,974 cattle were sent and 2,143 died. The East Delhi Municipal Corporation shared similar figures. Between August and October 2015, 190 cattle were sent to shelters. Of these, 120 died. Six to seven cows die every day. Most are old and diseased and some are nearly dead when they reach to Gaushalas.

Illegal dairies that operate from urban areas abandon those cows that are useless to them. A few let out milch cows to roam during the day and take them back in the evening. In 2002, the Supreme Court ordered authorities to rid Delhi's streets of cattle and shut all dairies functioning out of urban areas so that abandoned animals won't find their way to Delhi's busy streets. The order seems to have had little impact. Veterinarians routinely remove 20 to 30 kilograms of plastic after operating upon stray cattle- earning them the name plastic cows. Cow protection groups are also becoming a problem for cow shelter managers. They are ill-mannered, don't want to follow the rules and want shelter managers to take in cows even when they have no place. These incidents have increased over the years.

Roaming carelessly through crowded roads and eating whatever comes its way, with little or no restriction, the life of the holy cow in the national capital is more difficult than it looks. On the streets, cows eat garbage wrapped in polythene bags, used sanitary napkins, medical waste and nails. Many dairy owners inject a banned drug called oxytocin in cows for early onset of puberty and to prolong the milking age. The quality of milk produced by injected cows is harmful. For maximum produce, the calves are also not allowed to suckle on their mothers, which makes them weak and increases chances of contracting diseases and premature deaths. A glimpse of cow shelters and their working is given below (Source: HT -11.08.2016).

A Glimpse of Cow Shelters in Delhi (Source: HT 11.08.2016)

- 5 government-aided cow shelters in Delhi.
- 23,890 capacity of these cow shelters, 17,101 occupancy.
- 20 number of cattle die in these shelters per day.
- MCD helpline to report abandoned cattle: 1,266 in South and North Municipal Corporations and 1,55,303 in East Municipal Corporation.

Shree Krishna Goshala, Sultanpur Dabas, Delhi: The largest cow shelter, is also among the best maintained. The Gaushala has a full-fledged veterinary hospital, which has facilities such as a pathology lab, X-ray and ultrasound facilities and an operation theatre. It also has two ambulances to fetch distressed cows on calls. The cattle here are separated on the basis of breed, age and physical condition. This is done to maintain a comfortable living environment for them.

Acharya Sushil Gausadan, Gummanhera, Najafgarh: The most controversial because of issues of maintenance, a high cattle mortality rate and ownership squabbles, the Gummanhera shelter is everything a shelter ought not to be. Several dying bulls, cows and calves were found in shelter. There are 1,370 number of cattle, Capacity - 4,142, Area (over which it is spread) -19 acres. 55 cows dies at the centre in March 2016 due to poor hygienic condition and shortage of fodder.

Dabar Hare Krishna Gaushala, Surhera Village, Delhi: The cattle are brought here from all parts of the city and many cattle get severely injured en-route. Villagers are encouraged to donate for the upkeep, so that the cattle can be taken care of properly. Every morning the weak cows are examined by the doctors. There are facilities of performing surgeries in worse cases. There are 1,200 number of cattle, Capacity 5232, Area (over which it is spread) 24 acres.

Gopal Gausadan, Harevali, Delhi: The cow shelter, located on the Delhi-Haryana border, is crowded and also ranks low on maintenance. The cleanliness is a below par. Though there is a healthcare team, to take care of the veterinary need of the diseased cattle, there is no facility for conducting surgeries and other specialised treatment. Sheds are cleared once every day and all the government requirements are met. There are 3,976 number of cattle, Capacity 3,270, Area (over which it is spread) 15 acres.

Manav Gausadan, Rewla Khanpur, Najafgarh, Delhi: The cow sheds were reasonably clean but the place bore a derelict look. Funds are drying up and the money that the government provides is not enough to meet the day-to-day expenditure. There are 575 number of cattle, Capacity 3,488, Area (over which it is spread) 16 acres.

Pathetic condition of Gausadans in Delhi: According to the MCD's report on stray cattle drive, 95 percent of the cattle picked up by MCD and deported to the cow sheds or Gausadans die due to the inhuman and brutal methods used by the civic authorities. Describing Gausadans as slaughter houses, the report goes on to say that the MCD and Delhi government is not complying with the High Court order and has done nothing for their maintenance and welfare. A living cow in the Gausadans is a burden on the state exchequer but a dead cow is worth much more. Most of the deaths are unnatural and due to high degree of neglect by MCD.

Gausadans are basically meant for the terminally sick and aged cattle, the agency is even sending the healthy ones to Gausadans. It is brutal to pack healthy cattle in this space. Despite sufficient stock of medicine supplied by the Delhi government to all the sheds, these were not being administered to animals. Besides cattle brought by MCD are not being properly quarantined and vaccinated (Source: HT 12.07.2005).

Fund Constraints for Maintaining Gaushalas

Animal Welfare Board of India (AWBI) funded by the Environment Ministry - is the only central agency providing grants to organisations like the Prevention of Cruelty to Animals Act 1960. The bulk of these grants goes to Gaushalas with old, infertile and sick cows, besides those confiscated by the police while being transported for slaughter, in states where slaughter is banned.

An analysis of the ministry's budget allocations to AWBI shows drastic cuts in recent years, falling from ₹26.7 crore in 2012-13 to ₹11 crore in 2016-17. Consequently, the Board has given less to various animal welfare organisations.

There are about 4,000 Gaushalas in the country maintaining about 30-35 million cattle. The daily maintenance cost could be as high as ₹100 per cow. Ironically, this is way higher than the government's definition of the poverty line for humans, which is about ₹32 in rural areas and ₹47 in cities. According to an estimate, government funds account for barely ₹2 per cow annually, a drop in the ocean for Gaushalas, which largely depend on private charities. No efforts are on to make Guashalas economically self-sufficient. The funds released by Animal Welfare Board (₹crore) over the years are mentioned below (Source: TOI-26.06.2016).

Funds Released by Animal Welfare Board (Rs Crore)

Year	Total funds	Funds released to Gaushalas (%)
2007-08	12.8	7.7 (60.0%)
2008-09	22.8	14.8 (64.8%)
2009-10	22.0	14.4 (65.2%)
2010-11	20.2	12.9 (63.7%)

2011-12	21.8	14.6 (67.2%)
2012-13	16.1	10.2 (63.7%)
2013-14	13.0	9.0 (69.4%)
2014-15	12.0	8.0 (66.5%)

Government Gaushalas Running out of Space: All stray cattle found wandering the capital's streets are caught by the three municipal corporations and sent to five government-run gaushalas — two each in the North and South and one in the East. In 2016, the North civic body caught over 4,300 stray cattle, while the South corporation caught almost the same number from areas under their jurisdiction. On an average, the three MCDs pick up approximately 500 stray cattle per month. In May 2017, the North Delhi Municipal Corporation had impounded 105 cattle, the East civic body 300 and South 306.

No Space in Cow Shelters in Delhi (Updated: Jun 21, 2017): One can witness stray cattle blocking roads in different parts of the city, and in many cases even causing accidents. The problem could worsen further as five major cow shelters, aided by the Delhi government, have started denying admitting any more cattle for the past two months due to space crunch. The three corporations of Delhi —North, South and East — catch stray cattle and send it to one of the shelters.

The Ghuman Hera Gramin Gaushala in Najafgarh has maximum space for 2,500 cows, Sri Dabur Hare Krishna Gaushala in Suhreda village can accommodate 1,500-2,000 cows, Manav Gau Sadan Rewla Gaushala has space for 600 animals, Sri Krishna cattle shelter in Bawana can shelter 7,800 cows and Gopal Gau Sadan Harwali Guashala can accommodate 2,000 cows. In 2002, the Supreme Court ordered authorities to rid Delhi's streets of cattle and shut all dairies functioning out of urban areas so that abandoned animals do not find their way to Delhi's busy streets.

The space is an issue in all the shelters. The other problem is insufficient funds. The funds @ ₹20 per cow per day from the government and the same amount given by corporation are not enough. The funds too are mostly delayed. The cows in the shelters are brought in very bad condition as most of them eat plastics or are injured in an accident. Around 50% of the cows that are brought to Gaushalas are ill or injured. There is a need to control influx of cattle in Delhi. People leave their cattle roaming on streets and take it during nights so that they do not have to feed. There is a need to put a control over such acts so that the problem is solved at its root.

Stray Cattle

There are lakhs of stray cattle across India and they do not get proper care once they stop giving milk. Our main aim should be to ensure that in a few years from now, India doesn't have stray cattle any more. A plan be developed to ensure that even when cattle stop giving milk, there are people to take care of them. Cows that have stopped giving milk are routinely driven out to the streets, where they forage on garbage. There are over 5 million stray cattle in the country, according to the central government's 2012 livestock census. However, there are only around 1,850 registered cow shelters, apart from a few hundred unregistered ones, according to Animal Welfare Board of India (AWBI).These stray cattle or unwanted cows are basically by-products of the milk industry, which washes off its hands once these animals stop giving milk. An economic value has to be put to cows even when they stop giving milk to address the problem. Once cattle stop giving milk, they should not be allowed to go out of villages. Instead, every village has common land and that land should be used to house the stray cattle. And later, farmers can get cow manure from such shelters. This way, their dependence on fertilisers, which results in subsidy bill of thousands of crores of rupees, will reduce. Basically, stray cattle need to be looked from an economic point of view.

Stray Cattle Menace in Cities

Stray cattle roaming around freely in the cities have become a serious menace. Several lives are lost and many are injured due to accidents involving stray cattle. The root cause of the problem is the unplanned dairies, inside and around the city. The owners, after milking the cattle, leave them loose so that they can graze outside. Non-milking cattle is also left loose to save on their feed. The problems of impounding these cattle are many. First, the paucity of trained man power with the MC to catch and load them in a truck. Secondly, lack of Gaushalas with adequate space. Third, the tendency of the neighbouring towns/cities to catch their stray cattle and leave them outside near the next city. That is like sweeping your own house clean, and leaving the garbage in front of the next house. Fourth, feeding these cattle once caught, and fifth, the owner gets the cattle released with pulls and pressures. And last but not the least, lack of cooperation from the public.

The solution lies in relocating the dairies (centrally) outside the city. Let no one be allowed to keep cattle outside his house, and sell milk. It creates such a muck in the area, and is a nuisance to the people. The authorities have to create infrastructure for these dairies so that the owners can operate them. Thereafter, creating huge Gaushalas, at appropriate locations, which can house thousands of such cattle. Once impounded, these should be released only with a fine of minimum ₹5,000 per animal. The money, thus, raised can be used to maintain the cattle pounds and also pay for the cattle catchers. There should also be regular auctions of these cattle so that it fetches revenue, as also eases pressure on the pounds. The MC should take it as a project and implement with due sincerity. The cooperation from the people in rounding up the stray cattle will be most welcome.

Cow Protection Schemes

Gaushalas: Gaushalas are protective shelters for cows in India. Gaushalas to focus on treating cows well, because of their religious significance in Hinduism and consequent cultural sensitivity towards their welfare. Gaushala, a Sanskrit word ("Go" means cow and "Shala" means a shelter place: Go + Shala = shelter for cows), means the abode or sanctuary for cows, calves and oxen. Apart from providing sanctuary, the Gaushala also rescues the animals destined for illegal slaughter.

The first Gaurakshini sabha (cow protection society) was established in the Punjab in 1882. The movement spread rapidly all over North India and spread to Bengal, Bombay, Madras presidencies and other central provinces. The organisation rescued wandering cows and reclaimed them to groom them in places called Gaushalas. Charitable networks developed all through North India to collect rice from individuals, pool the contributions, and re-sell them to fund the Gaushalas. Signatures, up to 350,000 in some places, were collected to demand a ban on cow sacrifice. Between 1880 and 1893, hundreds of Gaushalas were opened. Between 2014 and 2016 the Indian government spent 5.8 billion rupees (US$87 million) on cow shelters.

Gokul Gram

Under this component it is proposed to establish Integrated Indigenous Cattle Centres or Gokul Grams in the breeding tracts of indigenous breeds. Gokul Grams will be established in: i) the native breeding tracts and ii) near metropolitan cities for housing the urban cattle. Gokul Gram will act as Centres for development of Indigenous Breeds and a dependable source for supply of high genetic breeding stock to the farmers in the breeding tract. The Gokul Gram will be self sustaining and will generate economic resources from sale of A2 milk, organic manure, vermi-composting, urine distillates, and production of electricity from bio gas for in house consumption and sale of animal products. The Gokul Gram will also function as state of the art in situ training centre for Farmers, Breeders and MAITRI's.

Each Gokul Gram will be set up by the Environmen Impact Assessment (EIA) and function under the auspices of the SIA/EIA or in a PPP mode. The Gokul Gram will maintain milch and unproductive animals in the ratio of 60:40 and will have the capacity to maintain about 1,000 animals. Nutritional requirements of the animals will be provided in the Gokul Gram through in house fodder production. Disease free status of Gokul Gram will be maintained through regular screening of animals for important diseases like Brucellosis, TB and JD. An inbuilt dispensary and AI centre will be an integral part of the Gokul Gram. Gokul Gram will also be set up near to metropolitan cities for managing urban cattle. Metropolitan Gokul Gram will focus on genetic upgradation of urban cattle.

Rashtriya Gokul Mission: To prevent unproductive cows being sent to the abattoir, the government started the Rashtriya Gokul Mission in mid-2014, a national program that involves constructing heavens for retired cows. Proceeds from the animals' bodily waste are intended to pay for their upkeep. In May 2016 the Indian national government held an inaugural national conference on Gaushalas. The State of Rajasthan has created a Ministry of Cow Affairs to advocate for animals.

The Rashtriya Gokul Mission is a focussed project under National Programme for Bovine Breeding and Dairy Development, an amount of ₹500 crore during the 12th Five Year Plan was allocated for development, preservation and conservation of indigenous breeds.

Objectives of Scheme

- To undertake breed improvement programme for indigenous cattle breeds so as to improve the genetic makeup and increase the stock.

- To enhance milk production and productivity of indigenous bovines.
- To upgrade nondescript cattle using elite indigenous breeds like Gir, Sahiwal, Rathi, Deoni, Tharparkar, Red Sindhi.
- To distribute disease free high genetic merit bulls of indigenous breeds for natural service

Components of the Scheme

- Establishment of Village Level Integrated Indigenous Cattle Centres viz "Gokul Gram":
- in the breeding tracts and ii) near metropolitan cities for housing the urban cattle.
- Strengthening of bull mother farms to conserve high genetic merit Indigenous Breeds.
- Establishment of Field Performance Recording (FPR) in the breeding tract.
- Assistance to Institutions/Institutes which are repositories of best germplasm.
- Implementation of Pedigree Selection Programme for the Indigenous Breeds with large population.
- "Gopalan Sangh": Establishment of Breeder"s Societies: Gopalan Sangh.
- Distribution of disease free high genetic merit bulls for natural service.
- Incentive to farmers maintaining elite animals of indigenous breeds.
- Heifer rearing programme.
- Award to Farmers ("Gopal Ratna") and Breeders" Societies ("Kamadhenu")
- Organisation of Milk Yield Competitions for indigenous breeds.

- Organisation of Training Programme for technical and non technical.

Implementation and Outlay: Scheme is proposed to be implemented on 100% grant-in-aid basis with an amount of ₹500 crore during the 12th Five Year Plan for implementation of mission namely „Rashtriya Gokul MIssion". Mission will be implemented with an allocation of ₹150.00 Crores during 2014-15.

Implementing Agency: Mission will be implemented through the "State Implementing Agency (SIA) viz Livestock Development Boards. State Gauseva Ayogs will be given the mandate to sponsor proposals to the SIA"s (LDB"s) and monitor implementation of the sponsored proposal. All Agencies having a role in indigenous cattle development will be the "Participating Agencies" like CFSPTI, CCBFs, ICAR, Universities, Colleges, NGO"s and Gaushalas with best germplasm.

The mission is to protect cow breeds. So, we can expect to see a rise in the number of local cow breeds. The government will spend a fund of 150 crore on the protection of the cows.

Many new Gaushalas may be constructed across the cities. If not Gaushalas, the government should construct cattle care centers. Let cow shelters be developed as training centers where people can learn about raising cows and making useful products from the animal's waste. The aim is to make people understand that cow is not simply important from religious point of view but a wealth in itself. Most people are not aware that we can earn good income by preparing some basic products, such as bio-gas and medicines using cow dung and urine. Cow learning is all about combining the religious aspect with the economical aspect attached with the cow. If a farmer takes the scheme seriously and helps the government to preserve these local cow breeds, the cow will get recognition. The government is planning to provide "Gopal Ratna" awards to such exemplary farmers.

Cow Sanctuary in MP (Kamdhenu Go-Abhyaranya): The country's first cow sanctuary, set up by the government, was inaugurated on 27 Sept. 2017 in Madhya Pradesh's Agar Malwa District. The sanctuary, also called Kamdhenu Gau Abhyaranya has been developed by the MP Gau Samvardhan Board in Salaria village, nearly 50 kms from the district headquarters. The sanctuary - Kamdhenu Go-Abhyaran- will be spread over 472.63 hectares and can house 5,000 cows in the first phase. The capacity will be increased by 1,000 every year. The second phase will focus on manufacture of organic fertilisers, improvements in the breed of cattle and preparation of medicines from cow urine. The third phase would involve the conservation of 34 breeds of cows found in India. Training would be offered in animal husbandry, preparing medicines from cow urine and developing grazing land.

7

Conserving Indigenous Cattle Breeds

There is a worrisome decline of indigenous cows. The Indian cow's downfall lies with decades of planned yet messy crossbreeding with a few bulls of even fewer exotic breeds like the Holstein-Friesians and Jerseys to boost milk yield. According to the 19th livestock census (between 1992 and 2012) exotic/crossbreeds (both milch and draught) have gone up by 14 percent, while the indigenous cattle have plumbed by the same percentage. Mechanisation of farming, which made draught cattle more or less redundant, and the growth of kerosene/LPG, which gradually replaced cow dung as staple fuel, pushed the native cattle further to the margins. As milk remained the only profit motive to own cattle, many small and marginal farmers too gradually switched to exotic crossbreeds. The descendants of the mythical Kamadhenu now contribute just over a fifth to the country's milk kitty.

India boasts of 30 pure cattle breeds, of which five, namely Sahiwal, Gir, Red Sindhi, Tharparkar and Rathi are renowned as milch breeds; Others as milch and draught cattle. The news about exotic crossbreeds is not good either.

None of the hybrids has been able to adapt to the tropical ambience. Indian exotic crossbreeds give, on an average, a paltry yield of 6.62 kg/day, while the Israeli Friesians boast of a prodigious 26 kg. Besides, as they require high maintenance, small and marginal farmers abandon them the moment the costs trump benefits. In Punjab, 80 percent of stray cattle are exotic crossbreeds, which are either left to die or sold as contraband to abattoirs. We should take inspiration from Brazilian farmers, who have been able to coax the pedigreed Indian breed Gir to yield an astonishing 56 kg/ day (Source: DTE 1-15 Nov, 2015).

The deterioration in the cow and its bullock is precisely because we have departed from the old practice of looking to the cow as giver of plenty. Even with all its deterioration, the Indian cow, through centuries of careful breeding, is in any way superior to its Western counterpart. The fat content of the milk of the Indian cow is rarely less than 4.5 per cent while milk from British cows has 3.5 per cent fat content only. Besides, the Indian cow can live on the meager fodder locally available, can resist disease and withstand the tropical heat much better than the British cow. The solution therefore to the question of improving the present condition of our cattle is not to cross the local cow with foreign breeds which, as a matter of fact, has proved disastrous, for the mixed breeds cannot stand the poor feeding and the climate, nor can they provide us bullocks capable of doing hard work. India has rich and diverse genetic resources, with some of the best breeds of dairy, draught and dual-purpose cattle. Despite the large number of good breeds, more than 80% of the cattle population belongs to the non-descript category. The important breeds are essentially the result of long-term natural selection and evolution over the centuries and are better adapted to withstand tropical diseases and perform reasonably well even with low inputs in terms of feed and fodder. Despite 30 recognised cattle breeds in India, an increase in the cross-bred population has been seen over the years.

Over 30 indigenous cattle breeds are essentially the result of long-term natural selection and evolution over the centuries and are fully adapted to different agro-eco-regions of the country, withstand tropical diseases and perform reasonably well even with low inputs in terms of feed and fodder. Indigenous cattle breeds are declining at a faster rate for example, in Andhra Pradesh, Ongole pure-bred population is decreasing and the Punganur breed has almost vanished. Similarly, Sahiwal population has dwindled in Haryana and Amritmahal in Karnataka. In Kerala, the Vechoor cow is almost on the verge of extinction. The possible reasons for the decline in the population are inadequate inputs in terms of feed/fodder, shelter and management, poor reproductive management, dilution of breeds due to inter-mixing, indiscriminate infusion of exotic germplasm for cross-breeding for enhancing milk production, absence of specific strategies and programmes for conservation of indigenous breeds and loss of breeds due to geographical re-organisation. Breeding tracts and organised farms of some of the established breeds such as Sahiwal, Red Sindhi and Tharparkar have been lost to Pakistan. On account of the prospects of quick economic gains, cross-breeding has encroached into even the native tracts of superior indigenous breeds, which has led to dilution of these breeds, some of which are threatened and heading toward extinction if no action is taken to arrest this trend. Even in the cross-bred strains, genetic stabilisation is yet to take place. The male calves of crossbred cattle are generally considered by the farmer as inferior as they are not useful as draught animals. These calves are neglected and starved to death, or more often than not, sold for slaughter for meat and hides, illegally in most states. These calves can not be slaughtered due to social and ethical reasons in few states of India.

India has a rich reservoir of genetic diversity and possesses some of the best breeds of cattle in the world. Our indigenous cattle breeds, such as Gir, Ongole (Nellore) are thought to be the world's oldest cattle. Zebu cattle bullocks were introduced in the United States as early as 1849. Brahman breeds in USA have been developed from Indian cattle breeds. First Indian cattle breeds (like Gir and Ongole) landed on Brazilian shores in 1906 are the finest Zebu specimens. The country's population of 218.8 million accounts for 17% of the total world population of cattle. The best indigenous germplasm of milch, draught and dual purpose animals accounts for 22-25% of the Indian cattle population, while 8-10% of the cattle population is cross-bred. Remaining (65-70%) are non-descript. Indigenous cattle breeds excelling in draught capacity, exhibiting a distinct superiority in utilizing poor quality feed and resistant for several tropical diseases are fully adapted to Indian conditions produce milk varying from 1800 to 3500 kg per lactation.

Decline in numbers of Indigenous Breeds

The indigenous breeds are declining in numbers in several states. For example, in Andhra Pradesh, Ongole pure-bred population is decreasing and the Punganur breed has almost vanished. Similarly, In Haryana Sahiwal population has dwindled and Amritmahal in Karnataka is also reported to be declining in numbers. In Kerala, the Vechoor cow was almost on the verge of extinction. The reasons given for the decline vary from indiscriminate cross-breeding to lack of grazing facilities, poor economics of maintaining some of the breeds and so on. States have answered in the affirmative. Some States like Andhra Pradesh have given suggestions for conservation and improvement for important breeds like Ongole.

A decline in population and productivity, dilution in purity of breeds and extinction of few indigenous breeds are serious issues and need immediate addressal. 13.79 M (6.4%)

cattle slaughtered every year results in negative growth in cattle population during 1992-97 (204.58 M in 1992 reduced to 195.58 M in 1997). Productivity does not commensurate with number of animals. Milk yields vary from 1800 to 3500 kg per lactation which is about half of the World average (2070 kg/animal/annum) and just 12-15 percent of the yields in USA, Canada and Israel. Indiscriminate breeding has led to dilution of indigenous breeds (Indigenous -22-25%, Cross-breds-8-10 % and Non-descripts-65-70%). Percentage of non - descript animals is increasing over the years and a few indigenous breeds eg. Red Sindhi and Tharparkar in Gujrat & Rajasthan, Ongole and Punganur in Andhra Pradesh, Sahiwal in Haryana, Amritmahal in Karnataka and Vechoor in Kerala are at the verge of extinction.

Breeding Policies and Strategies

The National Cattle Breeding Policy (NCBP) which has main focus on: 1) improvement of nationally important indigenous breeds through selective breeding in their home tracts, 2) improving low producing non-descript cattle by upgrading with other superior indigenous dairy breeds, 3) as an alternative to 2), low producing non-descript cattle can be cross-bred with exotic dairy breeds and 4) to continue inter-se breeding among cross-bred cattle using pedigreed/proven bulls, should be strictly followed. State Governments are required to develop region-specific breeding plan and strategies in consonance with the national breeding policy. Action Plan for implementing the breeding plan includes: 1) establishing an institutionalised monitoring mechanism from the Centre downwards, 2) drawing of region-specific and breed-specific breeding strategies in each state, 3) fixing targets in terms of actual numbers of cattle heads of the particular breed, infrastructure facilities such as sperm stations, bull farms etc. for each state, 4) organizing regular review meetings to assess the results in terms of the physical and financial targets, and to draw corrective action plan wherever the targets are not met and 5) ensuring adequate funding from Central Government to the State Governments to implement the programmes.

As regards Breeding Strategies, the Working Group felt that programmes for genetic improvement undertaken in the past were not successful, particularly those relying on up-gradation of indigenous breed through continuous crossbreeding for lack of backup support in the feed and fodder resources.

Since most of the female stock is needed for herd replacement, accurate selection of sires assumes greater importance. But a feasible cost effective and proven method for general adoption in the country is yet to emerge. Any programme for genetic improvement needs an organisational set-up that goes beyond the individual/herd. Absence of breeders' organisations and field recording network are serious handicaps in the emergence of viable and effective Breeding Service Organisations. Genetic measures undertaken to improve livestock will not be successful unless the livestock production system as a whole is considered.

The Working Group further states that the efforts would need greater attention because breeding is a cost intensive, long-term exercise with a time horizon of 15 years in India. Unless those who undertake such breeding programmes do not have a full control over various facets involved for this period, they run the risk of wasting time, effort and resources. If livestock development sector is to be successful, in terms of generating income to farmers, returns to government expenditures, and in value addition in international prices, the focus of policy will have to shift from the "best" technology to the most productive technology that is appropriate for different regions and is in tune with their natural endowment and labour and capital resources.

Cattle breeding policies and programmes in planning process: The issues related to cattle breeding have been dealt in the successive plan periods starting from the First Five Year Plan. The issue of surplus cattle has been handled in different ways in the various plan documents. Other issues such as cattle development, genetic improvement and breeding etc. have received varying degrees of emphasis in the different plans.

During the first Five Year Plan, a programme for improvement of cattle was suggested for production and use of adequate numbers of superior bulls of known parentage and productivity and elimination of inferior and unapproved bulls.

During the Second Five Year Plan, emphasis was given to increase the supply of milk, and to provide efficient bullock power for agricultural operations in every part of the country.

The Third Five Year Plan document took note of the seriousness of the problem of surplus and uneconomic cattle and arrived at the conclusion that weeding of inferior stock was a necessary complement to a programme of cattle improvement. With a view to eliminate scrub male stock, it was proposed to undertake a large-scale programme of castration during the Third Plan.

During the Fourth Five Year Plan, it was proposed that the schemes of the Third Plan including those relating to cattle breeding farms, bull rearing farms, Goshala development, control of wild and stray cattle and organisation of mass castration would continue and three central cattle breeding farms and eight bull rearing farms would be set up. It was also indicated that sire-evaluation cells would be established in each state.

In Fifth Five Year Plan, the emphasis was for continuing cross-breeding of cattle through establishment of exotic cattle breeding farms and intensive artificial insemination measures. Progressive introduction of artificial insemination technique using superior breeding bulls was the main plank for cattle development under the Key Village Scheme and the Intensive Cattle Development Programmes during the Sixth Five Year Plan. The document for the Seventh Plan period speaks for sincere efforts to increase productivity of milch cattle through the establishment of 500 Key Village Schemes and 122 Intensive Cattle Development Projects. Cross-breeding with exotic dairy breeds was accelerated through the establishment of frozen semen stations in different states.

During the Eighth Five Year Plan, emphasis was given to research in frontier areas such as genetic engineering for rapid upgradation of cattle through the use of embryo transfer and related technologies. By Ninth Five Year Plan, a considerable improvement in production of milk was noticed which was attributed to the intensified activities particularly, in improvement of genetic stocks, through cross-breeding, effective control of diseases and the Operation Flood Programmes. The major recommendations of the Working Group on Animal Husbandry set up by the Planning Commission for the Tenth Five Year Plan are: 1) A new programme focused exclusively on draught breeds of livestock may be initiated. 2) The National Project for Cattle and Buffalo Breeding be continued and the stipulations made for its implementation be followed in letter and spirit to realise the envisaged targets. 3) A well defined livestock breeding policy is in place which states that pure indigenous well defined breeds should be improved through selection, while non-descript low producing populations should be improved by grading up with other superior indigenous breeds or crossbreeding with exotic males. Indiscriminate breeding/crossbreeding with exotic breeds is threatening the very existence of these breeds in their home tracts.

The government should initiate steps to create incentives for breeding indigenous elite breeds and improve them through selection. 4) An aggressive strategy is to be adopted to remove the hurdles in sourcing and use of quality bulls for breeding. Military dairy farms could be used as a major source of crossbred bulls. 5) Monitoring cell for certification of sperm stations and A.I. bulls should be established in each state. Only certified semen should be used for A.I. 6) Rapid computerisation of the breeding network needs to be done in order to build up a reliable database and effective monitoring through a Management Information System (MIS) both at State and National level. 7) Advance technologies like ETT and OPU-IVF should be used to support breeding programme.

National Cattle Breeding Policy

On the basis of the recommendations of the National Agricultural Commission, a National Breeding Policy was evolved, which inter alia suggested:

(1) Development of nationally important indigenous breeds of cattle, both for draught and dual purpose through selective breeding in their home tracts.

(2) Cross-breeding of low producing non-descript cattle with exotic dairy breeds.

(3) Continue inter se breeding among cross-bred cattle using pedigreed/proven bulls.

State Governments were required to develop region-specific strategies and detailed district breeding plans, in consonance with the national breeding policy. The fact that there is a distinct depletion in the resources and number of recognised indigenous breeds, which now constitute perhaps only 20% of the total cattle population, may be even less, indicates that the respective State Governments have not been successful in conserving and preserving the good indigenous genetic germplams.

Development of Schemes/Infrastructure Facilities: For implementation of the recommendations of the Working

Group on Animal Husbandry set up by Planning Commission, a number of schemes (listed below) have been devised to achieve the envisaged targets of Cattle Breeding Policy in letter and Spirit.

Central Herd Registration Scheme

The Government of India, through the Department of Animal Husbandry, is running this Scheme, which envisages the registration of elite cow and buffaloe breeds of national importance. The objectives of the Central Herd Registration Scheme are: to locate superior germ plasm of indigenous breeds in breeding tracts, to introduce milk recording for further propagation, to regulate sale and purchase within the country and abroad and to propagate and awaken consciousness among the breeders for scientific breeding, development and preservation of cattle. The indigenous cattle breeds covered by the Scheme are Gir, Kankrej, Haryana and Ongole. The Scheme awards incentives for rearing of elite cows and male calves and provides a superior quality germplasm for superior breeding.

Cattle Breeding Farms

For rearing bull mothers of different breeds, five Central Cattle Breeding Farms (CCBF) were established at Suratgarh (Tharparker and its crosses with Holstein Friesian), Chiplima (Red Sindhi and its crosses with Jersey), Sunabeda (Jersey), Andeshnagar (Holstein Friesian X Tharparkar) and Hesserghatta (Holstein Friesian). The primary objective of these farms to produce at least 10 progeny-tested bulls in each farm was not achieved. The Working Group on Animal Husbandry for the Tenth Plan suggested for changing the goals of these farms and they should be used either for conservation of indigenous breeds which are at the verge of extinction or NDDB could take over these farms for implementing progeny testing programme using recent technologies like ETT/OPU-IVF.

Central Frozen Semen Production: Central Frozen Semen Production and Training Institute established in 1969 at Hesserghatta is a premier Institute producing above 9 lakh doses of semen per year and imparting training to the field officers and veterinarians.

National Project for Cattle and Buffalo Breeding (NPCBB): The Project came into being in October 2002 by merging two Centrally-Sponsored Plan Schemes, viz: i) Extension of Frozen Semen Technology and Progeny Testing Programme and ii) National Bull Production Programme. The salient features of the project are: i) to facilitate artificial insemination (AI) services at the farmer's doorstep, ii) to bring all breedable females among cattle and buffalo population under organised breeding plan within a period of 10 years, iii) to improve and preserve important indigenous breeds and iv) to train 14000 AI practitioners.

Recommendations of the Committee on Breed Improvement And Preservation: The following are the recommendations of the Committee on Breed Improvement and Preservation.

(1) **Breeding Policy:** From sixties onwards there have been directions to State Governments to take measures to conserve the important national breeds. At the same time cross-breeding has been recommended as a major tool for increasing milk production. The Committee makes the following recommendations in this regard:

- The Government should review its breeding policy and provide more emphasis to conserve indigenous breeds. If required, a separate policy for conservation of indigenous cattle breeds and their germplasm should be drawn up and translated into an implementable programme.

- Cross-breeding with exotic strains should be totally banned in the home tracts of the important cattle breeds and the ban should be strictly got implemented by the State Governments.
- A judicious mix of cross-breeding with exotic strains and preservation of indigenous germplasm should be maintained, while formulating the policy. Import of germplasm should be allowed only in very specific cases and after taking all the precautions to prevent the ingress of diseases into the country.

(2) **Implementation of the Breeding Programmes:** The broad outlines of the breeding policy are normally laid down for the State Governments to follow. While there has been a clause prescribing conservation of the important native breeds, especially in their home breeding tracts, no follow up or monitoring is being done at the field level. While the Central Government has been recommending, since the sixties, that steps and measures should be taken to conserve Indian indigenous breeds, this has not been translated into programmes and activities on a sufficient scale, by the State Governments. Some State Governments have prepared breeding strategies and regional plans, and have also identified the areas of breeding tracts for different breeds, on the ground these strategies have not actually translated into proper programmes and plans of action, which could achieve the desired results. The Committee recommends the following.

- A proper institutionalised monitoring mechanism be established from the Centre downwards, so that the implementation of the Breeding Policy directives is monitored closely.

- Directions should be given to the State Governments to draw up region-specific and breed-specific breeding strategies, programmes and plans to implement the conservation programme.
- Targets should be allocated to the concerned State Governments, in terms of actual numbers of cattleheads of the particular breed, infrastructure facilities such as sperm stations, bull farms etc.
- Regular review meetings should be held to consider the results achieved in terms of the physical and financial targets achieved, and corrective action taken wherever the targets are not met.
- The Centre in turn should provide adequate funding to the State Governments to implement the programmes and annual plans. Other sources of funding should also be tapped. (For example, the Haryana Government is reportedly collecting ₹0.10 per litre of milk from Gopalaks/milk producers and the money goes into a fund, which is used to supplement the efforts for breed improvement. The Government expects to collect ₹14 crore through this method).

(3) **Creation of Scope for Larger Use of Indigenous Cattle Breeds:** All the policy prescriptions and guidelines issued by the Centre had elements for conservation of indigenous breeds, their selective breeding for genetic improvement and use of indigenous breeds for upgrading low producing non-descript cattle in resource-poor areas. However, introduction of cross-breeding with exotic cattle even in such areas has seriously limited the scope of use of bulls of indigenous cattle breeds outside their breeding tract. The Committee recommends the following:

- The States should be directed to specifically delineate and identify, in their respective breeding policies, the geographical boundaries of the areas where non-descript cattle should be upgraded by crossing with bulls of indigenous breeds.

- Once such areas are earmarked, no cross-breeding of non-descript cattle, other than with bulls of indigenous breeds, should be permitted. This measure will provide an incentive to the farmers in the breeding tracts to rear male stock of indigenous breeds up to the breeding age, as the demand will create a market for the bull semen or natural service. Consequently, the practice of disposing off the male calves for slaughter will be curbed to a large extent.

(4) **Supply of Good Quality Breeding Material in the Breeding Tracts:** Due to, sometimes indiscriminate cross-breeding, there has been a general lack of emphasis on selective breeding in the breeding tracts. The supply of good quality germplasm, to match the quality of breeding females, has also been lacking. Conservation efforts have mostly been limited to institutional farms with small herd sizes, leaving larger parts of the breeding tract totally neglected. This cannot be really termed as conversion in its true sense and, more or less, amounts to only providing lip-service for the cause. The Committee's recommendations are:

- The status of the indigenous breeds needs to be evaluated afresh. This is not only because the composition of cattle in the breeding tract has changed, even the specimens and genetic make-up of the breeds have undergone changes over the past few decades.
- Breeds, which no longer find favour with the farmers, whatever the reason may be, should be identified and these breeds should be preserved only in the institutional farms, with improved conservation technologies.

- Breeds, which are accepted by the common farmer, should be developed through region-specific and breed-specific programmes, aimed at selection in the breeding tracts and supply of improved quality of germplasm for breeding of cattle for supply to farmers on demand.
- The progress of such programmes should be monitored through the institutional mechanism, recommended above to be set up.
- For sourcing cross-bred bulls, the Military dairy farms should be used as a major source of contribution to the Bull production programme.

(5) **Promotion of Breeders' Organisations:** Presently, the conservation of indigenous breeds is mainly in the domain of the Government and semi-Government agencies. Instances of owners of specimens of indigenous breeds forming an association to articulate the specific needs and demands for improvement of the particular breed are conspicuous by their absence. In contrast, breeders in other countries form strong proactive groups, which lobby for furthering their interests, as for example, the Brahman Breeders Association in USA. As Government moves towards privatisation in various sectors, cattle and livestock breeding should be encouraged in the private sector, with Government playing a facilitating role and providing infrastructural inputs, wherever required. The Committee's recommendations include:

- Government should encourage and promote the organisation and establishment of breed-specific associations to represent the requirements for development of particular indigenous breeds.

- Such Associations can then form a Federation at the apex level to take up issues with the Government either at the State or Central level.
- Government should accept the private sector players as partners in the efforts towards conservation of Indian breeds of cattle and achieve better results by involving them in a participatory manner.

(6) **Enhancing The Role of Voluntary Organisations-NGOs:** There are a large number of Goshalas and, to a lesser extent perhaps, Gosadans and Pinjrapoles, which have quite sizeable populations of excellent specimens of various indigenous breeds. However, efforts made by such agencies towards conservation, are more often than not, not based on scientific methodologies. The efforts are also diluted by efforts towards supporting old, diseased and destitute animals, mostly of non-descript breeds. At present the terminologies of Goshalas, Gosadans and Pinjrapoles are often loosely used and are inter-changeable at will. The Committee's suggestions/ recommendations are:

- An inventory of Goshalas/Gosadans/Pinjrapoles having good specimens of indigenous breeds of cattle should be drawn up, alongwith the details and numbers of cattle-heads.
- Such organisations should be designated with some appellation, which would distinguish them from other organisations maintaining other non-descript, aged or infirm cattle. For want of a better name, may be the term 'Gaushala' could be used, with other organisations, not having the indigenous specimens, being termed as Gosadans or Pinjrapoles only.
- Each such designated organisation should adopt only one or two breeds, depending on the strength and

composition of their herds, and segregate them from the other cattle, which they may like to continue to maintain as part of their animal welfare role. A specific breed-improvement/conservation programme should be drawn up for each designated organisation in consultation and collaboration with Government agencies.

- Such organisations can also participate in the Government-sponsored programmes for rearing of male calves from weaning to maturity, for breeding purposes. The male calves on becoming bulls can then be supplied to farmers and other clients in the breeding tracts for mating with breedable females and for upgrading non-descript breeds in other areas.
- The designated organisations should also be provided with scientific and technical inputs and training for genetic evaluation and selection of germplasm for breed improvement and upgradation programmes.

(7) **Use of Science and Technology:** Since the advent and introduction of crossbreeding in the country, most of the techniques and methodologies for breed improvement have been used to produce cross-bred cattle. The application of such technologies for propagation and improvement of indigenous breeds, is a relatively rare phenomenon, one of the reasons for which may be the reluctance of the farmers, owning these breeds to accept modern methods and techniques for reproduction. The Committee's recommendations are:

- Scientific and technological intervention in breeding programmes should be urgently taken up as a priority by the concerned Governmental agencies.
- Technologies such as artificial insemination, frozen semen production, progeny-testing, embryo transfer technology (ETT) should be used, after proper evaluation, wherever required, so that modern up-

dated scientific methods can be used to give a fillip to the programme for conservation, preservation and upgradation of breeds.

- The comparative advantages of Artificial Insemination and Natural Service, should be studied and the appropriate method should be adopted according to the specific needs, requirements and location of different areas.

(8) Monitoring cells for certification of sperm stations and bulls for frozen semen, should be established at the State levels and only certified semen should be used for AI, as suggested by the Working Group on Animal Husbandry set up for the Tenth Plan proposals.

(9) **Statistical Data Base:** Last, but not the least, the Committee would like to comment on the availability of data, or rather the lack of such availability, especially in the context of the breed improvement and conservation programme. The quinquennial Livestock Census, conducted by the Directorate of Economics and Statistics, Department of Agriculture and Cooperation suffers from all kinds of qualitative and quantitative problems, apart from lack of timeliness. The problems of lack of timely and quality data are further compounded by lack of perception of the data-collectors while categorising the breeds and quantifying their performances, particularly in the Integrated Sample Survey for Estimation of Production of Major Livestock Products, which is conducted by the Department of Animal Husbandry of the Central Government, through their counterparts in the States. For example, there is a tendency to club both non-descript and good indigenous cattle breeds into a single category of indigenous cattle, resulting in distortions while recording parameters such as average productivity etc. There is also a lack of perception about the maximum and average productivity of crossbred cattle in various production environments. This has created the general notion that all indigenous cattle are poor yielders of milk and the average cross-bred is superior. On the contrary, the limited information available goes to indicate that quite a sizeable number of dairy cattle of indigenous breeds show much higher productivity than the average cross-bred cattle. This is corroborated by the production records available for indigenous breeds at selected Gaushalas, military farms, as well as by field samples, which often indicate much higher productivity of indigenous cattle than is shown in the results of the Integrated Sample Survey for estimation of milk production. The Committee recommends the following:

- A reliable data base should be developed with regard to all the details of indigenous breeds, including their breeding tracts, numbers, characteristics, genetic make-up, germplasm, the institutional farms where they are being preserved and/or conserved and so on.
- Data bases should also be developed with regard to non-descript, as well as cross-bred cattle.
- A proper distinction should be made in nomenclature and classification of indigenous breeds, especially the recognised breeds, separating them from non-descript varieties. This classification should be communicated to the data collectors at the field level so that estimates of milk yield and other production data can be correctly collected.
- The Livestock Census must be conducted in a timely manner and, more importantly, the results compiled quickly. Results which come five years after the date of the census become meaningless for policy formulation and serves only as historical data. It is understood that

a decision has been taken in principle to transfer the Livestock Census work to the Department of Animal Husbandry, but the staff and other infrastructure are yet to be provided. If this is so, the Committee recommends that the following steps be taken for timely conduct of the transfer or there is a danger of the Census itself falling between two schools.

- While taking the Cattle census, the data-collecting agency should also gather details about the indigenous breeds, such as the name of the breed to which the specimen belongs, age, productivity etc.
- If it is not feasible to collect the detailed data through the Livestock Census, which is conducted by laymen, as is the population census, a special Cattle Census should be got conducted in all the States, especially in the major cattle populated areas and breeding tracts to collect all the details.
- The data of the Cattle Census should be fed into the data-base and then up-dated from time to time through surveys and other statistical methods for data collection.
- A breeding network should be set up by computerising and net-working all AI outlets, sperm stations, breeding farms and Goshalas and other agencies involved in the production of breeding material and implementation of breeding programmes. Activities such as performance recording, progeny testing, herd registration and sperm certification should all be computerised and connected with each other through the network.
- Monitoring of all aspects and facets of the breeding activities should also be done through use of the computerised mechanisms and networks.

Action Points to Conserve Cattle Breeds (Source: Agriculture Today-Sep 15, 2017)

- Establish Bull mother farms on priority for producing elite breeding bulls in public/private sector.
- Use of certified hygienic semen of progeny tested elite bulls for breeding of the animals/herds belonging to the farmers, government, private sector and goshalas in networking mode.
- Employ modern biotechnological tools like embryo transfer and marker assisted selection in superior elite animals of select breeds for specific regions for faster multiplication and desired sex.
- Provide suitable incentives to the breeders for conservation and genetic improvement of the breed.
- Reduce cattle population by producing more females by using sexed semen to save on feed and fodder.
- Ensure availability of certified quality semen and elite breeding males.
- Develop location specific cluster based strategies for each of the agro-ecological zones of the country. Local breeds of cattle should be conserved and further developed in their home tracts.

8

Improving Feed and Fodder Resources

Milk production in India is predominantly based on crop residues and natural herbage. Almost 95% of the total dry matter intake of cattle and buffaloes are in the form of straw, stovers, and on limited grazing on natural herbage. Unlike most of the developed countries, the production of milk based on land devoted exclusively for fodder crop production, is not feasible in India on account of the overwhelming demands of land to grow crops for human needs particularly food grains, oilseeds and pulses. Under these circumstances, production enhancement in cattle and buffalo has to be conceived on the basis of exploitation of the crop residues, chance grazing, marginally supplemented by green fodder and to lesser extent by balanced concentrate feed (Source: Agriculture Today-September 2017).

How to cope up with Fodder Shortage Particularly in Droughts

India faces a green fodder shortage of 63.5 per cent according to the vision document of the country's premier research institute - Indian Grassland and Fodder Research Institute

(IGFGRI). The shortage of dry fodder is 23.5 per cent. If the current situation continues then India's green fodder shortage will reach 66 per cent and dry fodder will reach 25 per cent by 2030 (Source: DTE-1-15 February, 2016).

During drought, cattle are the worst sufferers along with the people. They die of hunger, starvation and thirst. Farmers are unable to cultivate quick-growing fodder crop; without rain, grass does not grow. Cattle deaths, apart from causing losses to their owners, lead to the outbreak of epidemics. There was a time when large, common pastures were available for cattle grazing. All the cattle in the village were taken there. Gradually, with the increase in human population, the pastoral lands have been converted into agricultural lands. And attention has not been paid to earmarking land for common grazing. Panchayats, too, do not develop land under their possession for common grazing. Apart from fodder scarcity, there is a graver problem — water scarcity. The cattle may survive for a few days without fodder but not without water. Now, farmers and cattle breeders abandon their cattle rather than see them dying due to lack of fodder and water. There is a need to establish State and regional fodder banks to meet such crisis.

The reason behind the crisis is that nobody really cares about fodder in the country. Today's crisis is a result of decades of negligence. It is a side effect of our development process and government apathy, coupled with shrinking common land and lack of research work on fodder crop. The country has achieved self-sufficiency in food grain at the cost of fodder. The tradition fodder crops have been replaced by cereals and cash crops. Government data clearly shows rice and wheat have reduced the area under cultivation of traditional fodder crops such as barley, millets and other coarse cereals. Maise is the only fodder crop the cultivation area of which increased in the past 50 years because 60% of maise is used in the feed Industry.

There is need for a national fodder policy. Other needs are assured minimum support price for fodder crops especially coarse grains, fodder crop loans and attracting private investors for production of quality hybrid seeds of fodder. There is also need for a national grazing policy that will involve local governance institutions to manager common/pasture land.

In areas that regularly face droughts, hydroponics, a fodder variety that requires little water, grows quickly and takes little space should be promoted. Additionally, there are no seed standards for perennial grasses and legumes. Rampant use of uncertified, poor quality seed is adding to the problem. Usually farmers do not produce seed. And in the absence of standards, the seeds available in local market are of poor quality. A policy of collecting and preserving fodder for lean periods also needs support.

Interestingly, there is enough space to revive fodder in the country. India has nearly 25 million hectare of fallow land which can be easily used for production of fodder. Even if we cultivate fodder in 10 percent of the fellow land, the current deficit can easily be met. The revival of several natural grass varieties that would earlier grow in the wild-cenchres, heteropogon, chyrsopogon, cynodon dactylon and sawan grasses-can also help. Till 25 years ago, these wild grass varieties fed 70 to 80 per cent of the livestock population in rural India. They were completely rainfed and did not require any support from farmers. But encroachment of common land over the years has substantially reduced these fodder species. They were drought-tolerant and as a result provided fodder even during the lean summer season.

Production of Low Cost Feeds from Crop Residues and Biomass: Feed and fodder is one of the most important contributing factors for the growth of livestock sector. It is estimated that the 60-70% of total cost in livestock production is due to feed and fodder. At present the country faces a net deficit of 35.6% of green fodder, 26% of dry-crop residues and 41% of concentrate feed ingredients. As per the estimates, the deficit of dry fodder, concentrates and green fodder by 2020 is likely to be 11%, 35% and 45%. Adequate availability of livestock feed and fodder both quantitatively as well as qualitatively would be one of the key inputs in the growth of livestock sector in future, which can increase milk production by at least 20% with the current genetic resources available (Source : Agriculture today Sept 2017).

Production of low-cost feeds from crop residues and bio-mass, formulation of concentrate mixtures from locally available or leftover byproducts, efficient value addition and strategic supplementation of macro and micro-nutrients to available fodder resources is need of the hour. With the constant increase in cereal production, as a result of green revolution in the country, there has been a spillover effect, rather a matching or even greater increase in the availability of cereal straws (straw: grain ratio varying from 1:1 to 1:3 in different crops).

Judicious Use of Feed and Fodder Resources

Strategies and action plans for efficient use of available feed and fodder resources, securing surplus biomass for use in deficit areas and green fodder production enhancement from land available are mentioned below:

- Non-conventional crop residues such as cotton stalks, soybean and mustard straws, maise stover and cobs, pulse straws like pigeon pea straw and any other lignified biomass available in and around different milk sheds can be used for manufacturing straw based pellets/ blocks.
- Appropriate Mowers, Reapers, Liners, Bundle Makers, Balers, Tedders, Swathers, Choppers, Shredders, Threshers, Conditioners, Grinders, Mixers, Enrichment

Devices, Dosers, Material Handling Equipment need to be introduced and propagated, as per biomass needs.

- To minimise loss of crop residues Like paddy during or after harvesting, there is a need to upgrade/ modify available reaper, thresher and harvester machines for appropriate management of wet crop residues like paddy and propagate in different milk sheds.
- As area under fodder crops is static, there is a need to be promoted non-conventional fodders like moringa, thornless cactus, azolla, duaL purpose cereal crops to maximise land productivity without adverse effect on agriculture production and encouraging fodder cultivation on bunds/farm boundaries, canal side/national highway etc.
- To minimise seasonal variations in fodder availability, fodder conservation is to be propagated to conserve surplus fodder of maise, sorghum, pearl millet and oats through silage/hay making.
- Establishment of Micro Training Centre (MTC) by identifying progressive dairy farmer as a trainer may quickly disseminate improved technology to large number of users.
- Establishment of 'National Feed & Fodder Forum'. This Forum can meet once or twice in a year for monitoring feed and fodder situation in the country. The Forum may assist in implementation of Animal Feed Regulation Act at state level and publishing data on feed and fodder availability from a single source.
- Prioritise a special drive for R&D for feed and fodder resource inventory and development involving private sector for evolving better varieties/cultivars of forage crops. Production and availability of quality fodder seeds for rabi and kharif season for growing year round green fodder need to be taken with proper incentives.

- Remove VAT from feed and feed ingredients, electricity rates for small dairy farms and plants should be on par with agriculture, besides low rate of interest on loan (3-4%), at least 50 % subsidy on small dairy equipment, production and supply of area specific mineral mixtures, provision of micro nutrient fortified complete feed blocks and animal vaccines.
- Application of technology to produce large scale feed blocks, feed enzymes and other innovative feed resources, needs to be deployed. Effective implementation of the ration balancing programme of NDDB and Accelerated fodder development programme of the government can ensure better feed availability and improved nutrition.

Adopting the Technology of Urea Treatment of Straw: Out of the several chemicals tried for cleaving the lignocelluloses bonds in fibrous crop residues and improving their digestibility, urea treatment has emerged as the most effective one and is particularly more feasible under tropical situation.

Compressed Feed Block Technology

The compressed feed block technology has the potential to provide ready to eat complete ruminant diet during scarcity period, which are further aggravated by natural calamities like recurrent drought and floods in India. Crop residues such as straws (wheat/paddy), stovers (maise, sorghum etc.) and bagasse (sugarcane) are the major component of ruminant's diet in developing countries. Economic and balanced feeding is extremely important for optimum livestock production.

Urea and Mineral Supplements

Though rice straw has many limitations as a feed, it is no worse, in general, than other straws. Digestibility determined both *in vitro* and *in vivo* is as good as for other straws and in experiments both wheat and rice straw diets, properly supplemented, have equal body weight gains in growing stock. A diet of rice straw alone is not adequate even to maintain animals, which will lose weight if no supplementary protein is given. This is true of all straws and roughages containing less than about 4 percent protein, since a level of approximately 6 percent in the diet is needed to prevent weight loss. Alternatively, a supplement of urea at a rate of 1 percent of the straw feed will raise its nitrogen content to a level about equal with the protein supplement.

The need for supplementary calcium and phosphorus has already been indicated; from the results of various experiments about 10 g of supplementary calcium and 5 g of phosphorus would appear to be needed. In areas where specific information on possible trace mineral deficiencies is not available, it would be prudent to give a complete trace mineral supplement; 30 to 50 g of common salt per head/day should also be fed. These mineral supplements will be of use only if a nitrogen/protein supplement is provided.

Alkali Spray Treatment

The alkali spray treatment of straws has been shown to improve digestibility and intake. The straw is sprayed or sprinkled with a dilute solution of NaOH at the rate of 1 litre/kg, and the moist straw is immediately fed to animals. The optimum concentration of the NaOH solution appears to be about 5 percent where straw is to be fed with limited supplements, although where treated straw forms only about 50 percent of the diet, 7-8 percent appears to be better. If a smaller volume of a more concentrated alkali solution is applied with a pressure sprayer, the treated straw, along with supplements, may be pelleted.

The grinding and pelletting of poor quality roughages is known to increase intake. Whether the separate effects of alkali treatment and grinding and pelleting are additive has not been established. It seems likely that the high temperature and pressure combination to which treated straw is subjected in the pelleting process enhances the effectiveness of alkali.

The factory-scale processing of rice straw into pelleted complete diets containing 60 to 80 percent straw appears to have good prospects in Asia, where large numbers of milch animals are kept in or near cities and feed is transported from the surrounding country side.

Utilisation of Paddy Straw as Animal Feed

Feeding livestock appears at present to be the best use to be made of straws. In the tropics the subsequent fermentation of the resulting dung would further improve the recovery of energy and nitrogen from the straw. In many of the intensive rice-growing areas of Asia all straw is even now fed to livestock, though it is not efficiently used by them owing to the scarcity of supplementary feeds. Urea, minerals and alkali spray treatment could boost efficiency greatly. In other parts of Asia, in the Americas, and perhaps in some other areas as well, a large proportion of the rice straw produced is not fed to livestock. The possibilities of feeding it in any of the ways described above should be seriously considered, the most appropriate way in any particular situation depending on what other feeds are available and in what quantities.

According to the research work done on rice straw, the pretreatment of paddy straw should either be avoided or made simple and cheap so that the process of treatment is economically viable. There is a need for improvement of already existing strains of lignin degrading micro-organisms. The large scale treatment methods of feed production from paddy straw should be standardised and the end-product should be evaluated, so that no toxic end-products are present. In tropical zones in the world, ruminants depend on year-round grazing on natural pastures or the animals are fed with cut grass and crop residues. Most of these areas face seasonal dry periods in which the availability of pasture decreases and also its quality by a reduction in the content of digestible energy and nitrogen. Due to the fact that in these areas paddy straw is abundantly available from cultivating paddy, farmers offer rice straw as the main roughage source to their animals. In most of the developing countries, the fast increasing population leads to an even increasing demand for food grains for human consumption, which does not leave sufficient land for the cultivation of animal feed crops. According to an estimate of National Commission of Agriculture, there is a deficiency of concentrates and green fodders in India. Paddy straw is the stem part of the rice plant that remains in the paddy after the head has been harvested. It is generated at the tune of about 1.0-1.5 tonnes per tonne of paddy harvested. In past years, several studies have been reported on the physical and chemical characterisation and utilisation of rice straw as ruminant feed. In addition, numerous methods of physical, chemical and biological treatments have been investigated, including supplementation with other feed stuffs or components in order to improve the utilisation of paddy straw by ruminants. Paddy straw is usually fed untreated without supplements in spite of the fact that many methods for improved utilisation of rice straw have been developed and recommended. There are several reasons for farmers not to apply the already developed methods for improved utilisation of straw, such as physical, socio-economic conditions and practical reasons. In general, the use of rice straw as an animal feed as well as its treatment is always an economic decision. Ruminants by the virtue of microbial population present in their rumen can digest cellulose - a high molecular weight polymer of glucose in the native form. But in paddy straw cellulose occurs in association with hemi cellulose and lignin, a high molecular weight, three dimensional random polymer of phenyl propane alcohol. Chemically, lignin-carbohydrate bond forms metabolic blocks that generally limit the action of microbial hemi celluloses and celluloses. Unless the lignin is depolymerised, solubilised or removed, the cellulose and hemi cellulose cannot be utilised for their conversion into protein by micro-organisms. It has been estimated that one tonne of straw on burning release 3 kg particulate matter, 60 kg CO,1406 kg CO_2, 199 kg of ash and 2 kg SO_2. The large amount of rice straw as a by-product of the rice production is mainly used as a source of feed for ruminant livestock. Rice straw is rich in polysaccharides and has a high lignin and silica content.

Stubble Burning is not Wise (It's Wealth and Feed for Animals): Millions of tonnes (agricultural stubble) is burnt every year by farmers mostly in Northern India. Crop stubble is burnt in two seasons:1) September-end or October after the paddy harvest and 2) End of April-May following wheat harvest. Every year around October and early November, farmers harvest the Kharif crop and sow the rabi crop.

The annual world production of cereal straws and stovers is approximately 2,000 million tones, however, the energy contained in this vast bulk of material is on the whole poorly utilised and its nitrogen incompletely returned to the soil. With the rising prices of both energy and nitrogen fertiliser, interest is developing in more efficient ways of utilising straws, presently used mainly as livestock feed and as compost material (small amounts are used for the manufacture of paper and fiberboards). In the tropical and subtropical areas of the world almost all straw is fed to livestock, the resulting dung being widely used as fuel. With straw feeding, energy utilisation is relatively efficient, except for the residual energy in the indigestible matter which is wasted, but the introduction of dung fermentation in place of dung burning increases the proportion of straw nitrogen which is returned to the soil, thus improving the overall efficiency.

Straw burning is practised particularly where crop rotations do not allow sufficient time for ploughed-in straw to decompose, and where livestock enterprises that could use the straw have not been developed. Burning means, of course, a complete waste of organic matter, and this, in addition, pollutes the atmosphere.

Rice Straw

All straws are a poor livestock feed, and rice straw is no exception. It contains about 80 percent of substances which are potentially digestible and are therefore sources of energy, but actual digestibility by ruminants is only 45 to 50 percent. Furthermore, the amount an animal can eat is limited to less than 2 percent of body weight because of the slow rate at which it is fermented in the rumen. The net result is an energy intake which provides little or no surplus energy for growth, work or production. The most important consideration in obtaining more animal products from straw in the Asian setting is to improve digestibility and intake so that more energy is available for productive purposes. Protein supplements increase intake, while the alkali treatment of straws increases digestibility and usually voluntary intake as well.

Straws contain only 3 to 5 percent crude protein. Animals on an unsupplemented straw diet will usually not gain any weight and very often will actually lose weight. To obtain any production the straw must be supplemented, preferably with both nitrogen/protein and energy. For good growth on straw diets, a level of 8 to 10 percent protein is needed for young stock; this also improves consumption and thus increases energy intake.

The level of phosphorus in rice straw (0.02 to 0.16 percent) is less than the level of about 0.3 percent that animals need for growth and normal fertility. A level of about 0.4 percent of calcium in the diet is usually considered adequate for livestock, and many samples of rice straw have this amount, the range being from 0.25 to 0.55 percent. Nevertheless, many balance experiments with cattle fed rice straw have shown negative balances of calcium even when the calcium content of the straw was apparently adequate.

Sub-optimal cobalt contents in rice straws from Assam and Orissa have been reported and areas of copper deficiency are widespread in India and occur in a number of rice-growing areas. Thus the routine provision of supplements containing at least these two trace minerals would seem to be warranted.

Rice straw differs from other straws in having a higher content of silica (12-16 vs. 3-5 percent) and a lower content of lignin (6-7 vs. 10-12 percent). Whereas in all other straws lignin is the chief cause of low digestibility, in rice straw it is silica. Rice straw stems are more digestible than leaves because their silica content is lower; therefore the paddy crop should be cut as close to the ground as possible if the straw is to be fed to livestock. Thirty percent of rice straw silica is dissolved in the digestive tract, absorbed as silicic acid and excreted in the urine.

Rice straw also differs from other straws in having a high (1-2 percent) content of oxalates. These are broken down in the rumen to carbonates and bicarbonates, absorbed, and then excreted in the urine. The pH of water extracts of rice straw is about 8 and that of urine from stock fed rice straw as high as 9. The high oxalate content has been implicated in the greater need for calcium supplementation. Water washing removes 30-40 percent of the oxalates and substantially reduces urine pH and titratable alkalinity; calcium balance is also improved. Washing also removes adhering soil, which is considerable.

Feed Trash/Plastic to Cow for Blessings: Depending on the way you define trash, you could say that cattle have a long history eating trash in India. Cows have always been fed kitchen scraps, often presented as a religious offering in the mornings. Earlier and even now, the first chapatti (flat bread) of the day is offered to the cows as a blessing of prosperity, and the last goes to the dogs to appease the ghost world which dogs are associated with. But these offerings are often made in vegetable and fruit scraps, the ubiquitous bi products of traditional Hindu vegetarian cooking.

Plastic is being consumed by feral cattle which forage for food in trash heaps. When people conceal their food waste within plastic trash bags cattle consume the plastic along with the food. If a cow eats enough plastic bags their digestion suffers when the bags are collected in their rumen (the cow's stomachs) or the intestines causing the cattle to die of impaction or starvation - both painful ways to die. The Karuna Society, an animal welfare society in Andhra Pradesh, removed 40 kgs (88 pounds!) of plastic from a cow they autopsied is known to everybody. Just think of the thousands of tonnes of plastic that are being carried in the bellies of cattle all over India. This is a poor life for the cattle that have bestowed so many blessings on humanity - especially in India.

Plastic not only affects the wellbeing of cattle, but also the people who consume milk from cattle who are allowed to roam free during the day and return to their owners in the evenings, or feral cattle, especially males that end up as meat whether legally, in the two states that allow slaughter and sale of beef, or in the illegal meat trade. Polymers in plastic are a carcinogen (a cancer causing agent) when consumed in animals or humans and ingesting milk or meat from cattle that have ingested plastic, sometimes in astounding quantities, is extremely harmful for human health. According to a study the people living near the dumping site in India have been exposed to relatively high levels of contaminants like dioxins, possibly through intake of bovine milk.

Impact of Crop Residue Burning:

(1) Burning 1 tonne of straw accounts for loss of 5.5 kg of nitrogen, 2.3 kg of phosphorus, 25 kg of potassium and 1.2 kg of sulphur.

(2) Heat from burning straw penetrates into the soil up to 1cm; it may cause death of soil organisms like pests and reptiles which are otherwise beneficial for the crops.

(3) 1 tonne straw on burning releases 3 kg particulate matter, 60 kg of CO, 1,460 kg of CO2, 199 kg of ash and 2 kg of SO2.

(4) Some of the major air pollutants can affect our respiratory system. The elderly, the children and those suffering from chronic respiratory and cardiac problems are particularly at risk, a healthy person could also be affected. Anybody can suffer from upper respiratory tract infections, breathing problems including heart problems.

(5) Burning also leads to an immediate decline in the bacterial and fungal population in the top 2.5 cm of the soil. Repeated burning permanently diminishes the bacterial population by more than 50%, increasing farmers' dependence on fertilisers and pesticides. Burning stubble in Punjab alone is equivalent to wasting subsidies worth ₹500 to ₹1500 crore.

Reasons of Crop Residue Burning: Crop residue is usually set afire due to cost concerns and short time gap between summer and winter crops, besides lack of incentives and equipment to manually cut down the stubbles. The other reasons are:

- Getting subsidies for farm equipment that can plough back stubble into soil is difficult. Buying equipment is very expensive. Burning the stubble is a cheaper option. 2) It also helps prepare the fields quickly for the next crop. 3) The problem is acute in North India because if paddy is harvested by September-end or October, stubble has to be burnt for wheat sowing in late October-November.

Action Plan to Curb Stubble Burning: Around 20 million tones of straw is burnt in the field itself, which is worth hundreds of crores of Rupees. Proper management of this feed is urgently required. The NGT banned crop burning in 2015. Fine for burning in Haryana is ₹5,000 per acre and in Punjab ₹2,500 for less than 2.5 acres, up to 5 acres - ₹5,000. Implementing the order had been difficult. Famers set crop residue afire mainly because of cost concerns and the short gap between summer and winter crops. Lack of incentives and equipment to cut the stubble are other issues. Banning and penalizing farmers will not help because how will you identify which farmer is burning crop? It's difficult. Plus the farmer is burning it out of compulsion. The government can support some mechanisation at the village level for taking out the straw and then putting it to use by composting it. The farmer should see economic value in the exercise. Plans from different sources to curb stubble burning are mentioned below.

- The National Green Tribunal Appointed Committee in a meeting held in July 2017 decided that steps would be taken to make crop stubble profitable by creating a market for it so that farmers don't burn stubble. We can't just arrest farmers for burning crop residue. It's a livelihood issue for them. We need to create a market for crop residue so that farmers get an economic value for the stubble.
- Other measures include sensitizing farmers about the benefits of not burning stubble.
- Approaching manufacturers of harvester machines to provide attachments or modifications that can scatter the stubble across the fields is another way to curb burning.
- There is also need to look at methods for organic decomposition of stubble. Our primary focus should be to encourage farmers not to burn stubble through awareness campaigns.

- The Centre should support to buy machines that cut stubble from the base.
- According to a recent NAAS Policy Brief (Source : TOI-18.11.2017), In-situ solutions that can run the crop residue into a mulch layer over the seeds of the next wheat crop are the most cost effective, quickest to implement and sustainable solutions.
- Providing reapers and bailers to the farmers which can convert straw into feed blocks and thus save the feed from wastage and make it a marketable surplus for the farmer and add to his income.
- Encouraging farmers to use wheat and paddy straw and bagasse (sugarcane residue) for the production of ethanol. Farmers may earn up to ₹2,500 per ton.
- One promising technology is the Happy Seeder, a tractor mounted implement that acts as a no-till seeder. Coupled with a spreader, it deposits the mulched rice residue around the seed in one simple operation. The technology can increase profits of farmers by lowering equipment and labour costs and reducing fuel, fertiliser and herbicide requirements, while modestly increasing yields. It is a win-win for farmers and the environment as it improves soil health by retaining nutrients and saves water by up to 1.45 million litres per hectare through reduced evaporation and eliminating pre-sowing irrigation. It is most ideal for Punjab and Haryana where declining water tables are a serious concern.
- Farmers can also use certain inoculants (bacteria which help promote plant health) to aid in faster stubble management.
- Wheat and paddy straw can also be used as substrate for mushroom production, cardboard and paper making, briquetting and making other products.

- Mechanised harvesting and monoculture of wheat and paddy are also behind the problem of stubble burning. The cropping pattern should change. In Andhra Pradesh, bean gram and black gram are planted while rice stubble decomposes on its own. The government should also procure pulses and oil seeds which are suitable for north India.
- Pelletisation of farm produce and then use it as a source of fuel by power plants preferably NTPC is another way to curb stubble burning. There is a need to devise a proper mechanism for transporation of residue from farms to pelletisation plants and also about the incentives which could be provided to farmers to create interest.
- Farmers should be persuaded to stop burning stubble. And the farmers who pledge not to burn stubble should be given financial incentive by the government.

Worries of Farmers, Punjab in Particular: Farmers don't have money to dispose the stubble in an eco-friendly manner within a window of just 15 days between paddy harvest and wheat sowing (the next crop). Burning of stubble can't be stopped immediately. It may be done in phases. For every 80-100 acres paddy sown, machinery such as rotavator, paddy shredder, tractor, ploughs worth ₹15 lakh would be required. To manage stubble over 100 acres, 50 litres of diesel would involve additional cost. Atleast ₹1,100 crore is required to manage stubble in Punjab. The state government has been demanding ₹1,600 crore for permanent solution to the problem. In the absence of funds the government has shown helplessness to deal with the problem. The Punjab government is, however, pinning its hope on biomass plants, which could consume around 4 lakh tonnes of paddy stubble. Till date, around seven such power plants are operational with a capacity of 62.5 megawatts. Stubble burning is because of use of combine harvester that cuts crop about 2 feet above the ground level. With ploughing crop is cut from the ground. Use of mechanisation has polluted air, water besides soil fertility. It has led to acute shortage of animal fodder also. Unemployment has increased. As a result you will find animals particularly cattle roaming on roads, streets and eating on garbage heaps. In other words, mechanisation did lot of harm not only to environment but to farmers. Now the agriculture has become a profession of loss to the farmers. Farmers are under debt and committing suicide. Finally use of bullocks by marginal farmers need to be promoted.

Stubble Burning - (Combination of SMS & Happy Seeder) A Way Forward: According to estimates by the international non-profit research body, Borlaug Institute for South Asia (BISA), the use of Happy Seeders increased almost four times this sowing season (2017) over the previous year (2016), with an estimated 75,000 hectares coming under it in Punjab and Haryana.

The combination of straw management system (SMS) and Happy Seeder is a win-win for farmers and the environment. It does away with stubble burning while increasing farmers' profitability and promoting climate smart agriculture. The spinoffs are sufficient - lower chemical load on soil, more crop per drop, reduction in CO2 emissions, better soil health, less weeds and sturdier plants with deeper roots.

Though there is a subsidy of ₹50,000 for SMS and Happy Seeder machines, these implements still cost over ₹1 lakh each. Experts say these technologies including the reversible plough (for potato and vegetable farming after rice crop), need to be further incentivised and a major push needs to come from the Centre and state governments well ahead of next year's burning season (i.e. 2018).

9

Scientific Approach of Cattle Keeping to Improve Productivity and Profitability

Improvement in productivity is the key to protect cattle wealth. Potentials are great to improve productivity of indigenous cows. In an American study with dairy cattle, it was estimated that nearly $300 million could be annually saved with scientific approach of animal keeping. Breakup of savings is: $135 million by reducing the average calving interval, $ 85 million by reducing calf mortality from 10 percent to 4 percent, $57 million by reducing the number of steriled culled cows from 5 percent to 2 percent and $ 20 million by reducing the number of services per conception from 2 to 1.5. The genetic impact of AI in the cattle industry is well established. In USA, by 1984, the milk produced from 20 million cows was equivalent to milk produced by 27 million cows in 1950. Impact of AI is visible in India too. Sexing of sperm and/or embryos will reduce the number of male calves born in dairy industry and will increase the number of male calves born in meat industry. The transfer of in-vitro fertilised, frozen, sexed embryos into oestrous synchronised animals will reduce the cost of breeding and result in more genetic gain. Cloning provides an opportunity for the production of large numbers of genetically identical animals descended from a superior animal. This system enables phenotypic selection and can be used to bring about rapid changes in commercially important traits, such as meat and milk production. The primary impacts of these research developments described above will be in improving lactational performance, improving growth and carcass characteristics, improving reproductive efficiency and increasing disease resistance. Ultimately, the effects of these advances will be to increase the efficiency, reduce the cost of production and improve the nutritional quality of major animal products. All these gains are indirect instruments which prompt farmers to conserve and safeguards their animals. Development of breeding policies and breed conservation strategies, research thrusts and an institutional mechanism are important to protect the indigenous breeds.

Wealth from Carcass Loses: India is losing nearly ₹600 crore per annum due to the non-recovery of material from fallen carcasses of buffaloes, bulls, goats and other animals. Most of the time the animals, who die a natural death, are just left to rot in the open for scavenging by vultures and other predatory animals. The disposal of these animals after a natural death falls upon the most disadvantaged sections of the society, who sometimes make meager earnings by selling the raw hide in the local weekly market to the local traders, who in turn sells it to a tanner. They generally make no attempts to recover any byproducts like meat meal and tallow left in the intestines, flesh and other organs at the outskirts of villages where they carried out their flaying exercise except perhaps bones which many of them collected later to be sold to the "bone-mills".

The need is for immediate collection of dead animals to stop rotting; provision of a transport vehicle (jeep and trailor for collection of carcasses; provision of flaying tools so that flaying could be done on scientific lines; developing indigenous equipment for recovering tallow and meat meal from the flesh and internal organs; treatment of effluents; training of flayers and other personnel in correct methods of carcass recovery and by-product utilisation. There is also need to build proper linkages between raw materials processing, product development and manufacturing besides market identification to boost export of finished leather and leather goods.

Recycle Slaughterhouse Waste into Organic Fertiliser: The waste generated in small abattoirs if recycled can provide additional income to slaughter house owners. A research group from Jadavpur university has been converting waste produced at slaughterhouses into organic fertiliser, which is particularly beneficial for growing tomatoes. Local farmers have accepted this new fertiliser as substitute for the currently used chemical fertilisers. The fertiliser has proven safe in terms of both pathogen content and heavy metals.

Processing and Value Addition

- Establish specialised mandies/kisan bazaars for milk and milk products supported by cold chain, primary processing and packaging at tehsil/block level.
- Introduce MSP for milk and milk products.
- Scientific processing of livestock waste generated at farm units, dairy farms, dairy plants, tanneries, feed mills, etc. to convert into biogas, compost, vermin-compost etc.
- Update farmers on useful livestock technologies, best practices and innovations; weather, market intelligence and input availability through mobile phone service, kiosks, community radio programs, print and multimedia, and training.

- Undertake value addition of milk in terms of new products following HACCP for improved keeping quality, long shelf life, and nutritional value.

Development of Infrastructure and Capacity Building

- Cover all the districts in the country by the Livestock Insurance Scheme through the Central/state schemes.
- Develop capacity building for testing the quality of feed, medicines, fodder seeds, residues in livestock products; production of sufficient doses of disease free semen, chicks, fingerlings, vaccines, diagnostics, cold chain, terminal markets for livestock produce, delivery of veterinary services at farmers' door step.
- Create adequate quality control labs for testing of food quality, vaccines, imported livestock and poultry, along with quarantine facilities.

Training and Extension Services: Extension services need sea change for effective and timely 24×7, deliveries at farmer's door step.

Frame Policy for Aged Cows: The government to frame a policy on how poor farmers should handle cows and buffaloes when they get old and stop milching.

Health Coverage: Currently due to lack of adequately trained manpower and their mobility, the services provided do not create desired impact. An authentic, concurrently updated database for prevalence and emergence of diseases is essential for identification, onward prevention and control. Infrastructure of vaccine and diagnostic production units, semen stations and AI breeding farms that is largely owned by the government can be more efficiently utilised by way of appropriate participation of the private sector. Provide mobile veterinary services for effective door step delivery. Each district should have at least one polyclinic. Disease investigation and forecasting system should be strengthened. Also provide veterinary services, such as AI, deworming, vaccination etc. at farmers door, through mobile clinics, training of Pashu Mitras/SHGs at village level by involving women, unemployed youth, along with strengthening of infrastructure in veterinary hospitals/dispensaries.

Application of Biotechnology to Improve Livestock Productivity: Biotechnology has the potential to produce more food from animals without increasing the animal population or exerting pressure on the eco-system and environment. Biotechnology is application of biological principles that influence animal breeding and reproduction, animal health (survival) and nutrition. Animal breeding and reproductive technologies are used for selection and faster multiplication of genetically superior animal germplasm, adoption of improved animal management practices, animal health care and milk procurement, processing and marketing.

Reproduction Tools To Improve Productivity: During the last two decades, farm animal reproduction has entered into an era of a new biotechnology revolution which includes artificial insemination, induction and synchronisation of oestrus (manipulation of breeding cycle), superovulation, embryo transfer, cryo-preservation, embryo resource development, sexing, transgenesis, cloning, chimera production and transgenesis. Some of these technologies have immense potential to revolutionise world animal agriculture in the twenty first century.

Artificial Insemination

At present there are 51 semen stations in India with a production capacity of 81 million doses/year against the current demand of 100 million doses for bovine semen and 150 million doses in the next few years. Most of the semen stations cater to the demand for buffalo semen and germplasm of exotic and crossbred cattle. The country needs to augment trained manpower including veterinary personnel to provide quick service delivery. Besides, provide quality equipment and appropriate training to avoid artificially inseminated cows becoming infertile and developing infections. There is need to upgrade local breeds using elite indigenous stocks such as Gujarat's Gir, Rajasthan's Sahiwal and Rathi, Deoni, Tharparkar and Red Sindhi of Rajasthan and Uttar Pradesh using AI and IVF-ET techniques.

Embryo Transfer

Application of embryo transfer includes its use in genetic improvement through selection of parents, multiplication and mechanical manipulation of superior germplasm using genetically unreliable mothers as foster mothers for embryos of superior genetic makeup. Successful embryo transfer involves several steps, viz: selection of donor parents and recipients, superovulation of donors, breeding of donors, synchronisation of oestrus in recipients, embryo collection, embryo evaluation, embryo processing, embryo storage, embryo transfer, post-transfer care of recipients and early detection of pregnancy. Failure to comply strictly with any one of these steps will lead to disastrous results.

Induction and Synchonisation of Oestrus

For satisfactory pregnancy rate in an embryo transfer programme, the embryo must be placed in an environment that simulate the one from which it is removed. Therefore, synchronisation of oestrus in donor and recipient in each set of experiment is considered essential to obtain the best results. Two alternate approaches are used for induction and synchronisation of oestrus. The first is by artificially extending the luteal phase (using progestational compounds) and second by inducing the demise of the corpus luteum (using prostaglandins and their analogues). Progestogen and prostaglandin in combination are also used. During the non-breeding and anoestrus season, treatment of the females to induce oestrus and ovulation is similar to that described for the breeding season. Progestogens with some possible adjustment in PMSG and/or FSH are used for anticipated lower response. PGF_2 alpha in double dose schedule (10-11 days interval) can be used to cover those animals who do not have functional CL at first injection.

Cryo-Preservation

Cryo-preservation of embryos has become a routine procedure for success of embryo transfer in livestock species. The methodology is used to freeze and store surplus embryos recovered from elite animals. The methology is also used to freeze and store oocytes for in-vitro fertilisation at a later date. Cryo-preservation of embryos has multifarious application, viz: relieves for having simultaneous synchronisation of oestrus in donor and recipient, easy and safe transport of valuable germplasm around the World, conservation of superior genetic material, protection of valuable strains of experimental animals against possible loss through disease, accident or genetic drift, possibility of shortening the generation interval for progeny testing programme.

IVF to Revive Desi Milch Cattle Breads

IVF-ET technology allows scientists to extract eggs from a high-yielding cow, combine them with semen from a bull of the same breed and transfer the lab-developed embryos into surrogate cows of other breeds.

Sex Selection to Produce More Cows than Bulls

According to the agriculture ministry, there are 300 million bovines in the country. Of these, only 85 million give milk while the rest-including 84 million males are "unproductive". Through the sexed semen programme, it is possible to add six million cows of a superior variety to the milch herd every year.

A sex ratio favouring males is more desirable in meat breeds because of the faster growth and lean carcass and females in dairy breeds. There are two approaches for the control of sex of offspring's: 1) regulation of sex of embryos and 2) identification of sex of embryos. Regulation of sex of embryos are based on i) sperm separation, ii) parthenogenesis iii) nuclear transplantation and iv) removal of pronucleus. Identification of sex of embryo involves removal of trophoblastic cells by biopsy/bisection of embryos. The sex of the biopsied cells is determined by i) karyotyping, ii) identification of sex chromatin, iii) assay of sex-linked enzymes, glucose - 6 phosphate dehydrogenase, iv) identification by DNA probes or Y chromosomes and v) serological determination of H-Y antigen. The target of sex sorted semen doses from superior sires should be enhanced to increase availability of high genetic merit heifers.

Bulls have become a burden on farmers because growing mechanisation in agriculture has made them near-redundant in the fields. They are often abandoned on the streets, where they loiter and eventually starve to death. The environment ministry's recent rules banning the sale of cattle for slaughter at cattle markets will only aggravate the problem (Source: HT 10.07.2017). Other countries get rid of unproductive male calves by culling them for meat. However, this is not an option in many states of India. Establishment of sexed semen production facilities will benefit Indian farmers to a great degree. "If farmers are able to ensure the birth of more females, it will boost milk production and agricultural income.

According to the Union Government Plan, 10 animal husbandry centres are to establish sexed semen production facilities to boost milk productivity and tackle the problems of "unwanted male calves" arising from cow slaughter bans in many states of the country. Sexing technologies allow farmers to control the gender of the calf. The use of sexed semen may result in the birth of nine females to every male.

Cloning: The ability to produce multiple copies (clones) of outstanding animals has potential significance to the livestock industry. Clones can be produced by 1) embryo splitting, 2) blastomere separation and 3) nuclear transfer. Identical twins have been produced using approach 1 and 2 but production of more than 2 clones is difficult due to low cell numbers at the blastocyst stage. Third approach is through transfer of embryonic cell nuclei. In this approach, the single blastomere of preimplantation embryo is fused with the zygote whose pronucleus has been removed. The other approach is the somatic cell nuclear transfer. A large number of clones can be produced after nuclear transfer of various somatic cells from mammary gland, cumulus, oviduct, skin, ear, muscle, liver, tail or sertoli cells during last 4-5 years.

Market Platform to Cattle Breeds and Germplasm: A web portal, a digital market platform to cattle breeders, sellers and buyers was launched on 27th November, 2016. The portal - www.epashuhaat.gov.in will provide a digital market platform to breeders, sellers and buyers to facilitate trade of cows/buffaloes, frozen semen and embryos under a verified and certified system without intervention of middleman. The features of web portal are mentioned below (Source: TOI-28.11.2016).

(1) The portal has connected farmers to 56 semen station across 20 states in the country. Besides being a buyer/seller platform, this government portal will also provide many related services to farmers and breeders.

(2) These services include handling of artificial insemination, good care practices, animal certification, nutritional advisory and disease screening and testing of animals.

(3) The portal is part of the ministry's initiative to meet the ever-growing demand of milk in the country under a new

scheme-National Mission on Bovine Productivity - with an allocation of ₹825 crore. It is also aimed at making milk production more remunerative to farmers.

(4) Though indigenous breeds are currently being sold in local 'haats' and animal fairs in different parts of the country, it does not guarantee any breed in absence of a verification mechanism. The ministry's portal also help the buyers in getting authentic breeds of animals and their scientifically verified semen and embryos.

10

Way Forward to Revive Cattle Wealth

Since time immemorial, India has been an agrarian country and cow has been its backbone. When fertilisers and tractors were unknown, cow was the only source of sustaining entire agriculture. Without cows, agriculture wouldn't have been possible. Cows provided the source of fertiliser (Cow Dung Compost) while bulls ploughed the land. Additionally cows provided milk and milk based products, which fetched good commercial value.

For last few decades, especially since Green Revolution and usage of chemical fertilisers, pesticides and tractors, have dealt a massive blow to importance of cows in agriculture. While productivity levels have improved in short term, it has damaged the farming ecosystem. The damage has been so extensive that now it requires a major overhaul, for which cow based organic farming is the only remedy. Not only the artificial chemical fertilisers are damaging the soil and water, but they are also very expensive as compared to cow based natural fertilisers. Due to this dangerous drive for usage of high productivity and chemical fertilisers, farmers have been lured into believing that they don't need cows any more. Which has accentuated the agrarian crisis leading to literally thousands of farmer suicides and banks in distress who lend money to them. To make cow based agriculture model economically viable, let us look at some cow products as mentioned below.

Cow Milk

Indian cow milk contains A2 type protein making it superior to exotic breeds. Due to this desi cow milk is in high demand and fetches good money to farmers. Milk products like ghee, paneer, butter, curd, etc. also fetch good amount to the farmers.

Cow Dung

Compost & fertilisers made from cow dung are organic, natural & cheap (Close to Zero investment) sources to increase crop yield for farmers. Farmers can save money by developing a Gobar Gas Plant. They can earn money by suppling cow dung cakes, another product in high demand in urban areas.

Cow Urine

Cow urine is like Amrit for farmers. Farmers can sell cow urine to many Ayurvedic companies where it is extensively used for manufacturing of medicines. They can save huge money spent on purchase of chemical pesticides by using insect repellent produced with cow urine.

Medicines

Medicines prepared from cow based products are excellent alternatives to costly chemical based products floated in market by big companies. Farmers can limit their spend by producing these medicines at almost zero cost at home.

With these accomplishments, the cow based farming maintains ecological balance beautifully. The fertility of soil increases with the use of cow based products. This makes it possible for farmers to take multiple crops in an year and earn more profits. There is need to spread awareness about cow based lifestyle. Arranging a training programmes to promote this concept may reduce input cost and increase farmers income. The irrational, non-economic and mythical aspects of the Indian cattle complex are greatly overemphasized. Rational, economic and historic interpretations of the taboos, customs and rituals associated with management of Indian cattle can lead to effective answers. The cattle wealth in India has to be examined under the adaptive processes of the ecological system, rather than Hindu theology.

Economics, Not Religion to Protect Cow Wealth: India's secular idea rests on the foundational principle that India belongs equally to people of diverse faiths and belief systems, cultures and languages. It prohibits the religious, cultural, legai or moral dominance of any one religion or culture over any other. Unlike many European countries that require residents from distinct cultures and faith-systems to absorb and adapt themselves to the dominant language, culture and practices of the country, the essential secular idea of India does not require any kind of conformity of faith, culture, diet, dress or language to qualify as a fully-equal Indian citizen.

It is this uniquely Indian secular idea that Hindu nationalists challenge, by demanding that non-Hindus conform to what they claim to be the majority faith. There is no doubt that many Hindus revere the cow. The campaigns against cow slaughter that we witness are not about love for the cow, but hate for fellow Indians. And a chillingly similar horrific mob lynching 12 years earlier in Jhajjar in Haryana of five Dalit men on the charge of killing a cow are a reminder that not all Hindus worship the cow (Source : HT 06.10.2015).

India, to become globally competitive, will need to create an efficient supply chain network through significant investment in infrastructure and human resources. R&D efforts need to be intensified for sustaining the cost of production, increasing milk yield per animal and its quality, fodder productivity and risk mitigation. This requires developing innovative farming models and motivating a large number of small milk producers to adopt them. Linking the production system to the consumer demand and processing units requires a robust value chain, intensive research and technology infusion. These strengths can further be leveraged with balanced growth of crop and livestock farming system. Strategic action plan should, therefore, focus on following aspects.

Equal Status: Dairy farming now deserves to be given equal status on par with agriculture, rather than its subsidiary status, in view of its share in agricultural GDP and employment. Village-level milk producing units should be brought in the organised sector and promoted in a systematic manner to convert existing individual sustenance dairy farms and traditional family farms into collective, community & commercial farms operating as business farms. This would call for intensive training and capacity building of dairy farmers [human resources] with focus on business-like operations, financial and marketing management.

Data Base: India should have a robust system to periodically generate survey based data & information on critical aspects of dairy sector that can help formulate five year perspective plan for targeted milk production state-wise bridging the gaps in institutional & physical infrastructure. India should develop a system to compete with developed countries where every milch animal is tagged with a number and every drop of milk processed, value added, marketed and instantly recorded.

Safety Standards: Unfortunately, a large share of the supply or production still does not conform to the domestic, leave alone global food safety standards. This is due to adulteration, lack of awareness and rigorous enforcement of Food Safety Standards and inadequate infrastructure comprising technology & trained manpower.

Hygienic Milk Production

68.4% of milk in the country did not conform to standards laid down by the food regulator. This may be one of the key reasons for India's 0.4% share in global export despite India accounting for 17% of the global milk output. To remedy the situation, it requires significant investments in terms of purchasing electronic milk-testing machines, electronic weighing systems as well as chilling and transportation equipment. If this is accomplished, the quality of milk collected directly from farmers will be better, unadulterated and relatively cheaper. The environment under which milk is produced, collected, transported, processed and distributed should be fully conducive and that animal raising practices related to sanitation, quality of drinking water, feed and fodder, type and quality of pipelines etc. must be aligned to the goal of healthy milk. Farmers need to be trained/guided to display high degree of hygiene and know-how of animal health care and nutrition.

Capitalising the Strength of Cooperatives

Dairy cooperatives as the people's institutions are the result of dairy farmers' entrepreneurship to exploit the potential of dairy markets in India. These institutions have to be driven like business enterprises combining professional management with technical and financial expertise. The need is to nurture dairy entrepreneurs through effective training of rural youths at the village level coupled with dedicated leadership and professional management of farmers' institutions/organisations.

Market Access

Dairy industry's potential for inclusive/equitable growth and income distribution in villages can be harnessed by enhancing market access and offering stable and remunerative prices to farmers. Integrating dairy and crop farming with value chain system can be a better source of sustainable livelihood for rural poor and most vulnerable families.

Export

India contributes about 17% of the global milk output but its share in global export is insignificant at 0.4%. A large quantity of milk still remains unprocessed. India is surrounded by countries and regions that are milk-deficient viz. Bangladesh, China, Singapore, Thailand, Malaysia, Philippines, Japan, the UAE, Oman and other gulf countries all of which are located very close to India. India, therefore, needs to have a systematic research & feasibility studies under Public-Private-Partnership mode to explore these hitherto unexploited international markets and initiate specific policy & programs on lines of Agricultural Products Export Development Authority in consultation with Union Commerce Ministry, Indian Institute of Foreign Trade and Indian Embassies in these countries.

Indigenous Cow Breeds to be explored for Their Potential: There is a misconception that Indian breeds do not produce milk in quantities comparable to foreign ones like the Jersey or the Swiss cows. Indian breeds such as Gir, Sahiwal are far superior.

Promoting Vegetarianism Rather Than Imposing Ban on Cow Slaughter: There are several things that cause the mind to become agitated, and violence is one of them. When we kill an animal for food, we not only eat the animal's flesh, but also take in its emotions and energy. The fear and panic that the dying animal experiences get absorbed by our energy bodies and this has a negative impact on our mental state. Our mind becomes agitated and violent thoughts enter our mind; once they take root, it's difficult to get rid of them. Therefore, it is advisable to avoid nonvegetarian food. Eating meat increases 'Tamo Guna' or inertia in the body; it then leads to laziness, anger, irritability, pessimism, criminal tendencies and other negative qualities. The criptures lay emphasis on eating 'Saatvik Food', which means eating vegetarian food. It is important to strengthen the saatvik nature through self-discipline and eating food that promotes only saatvik qualities. Vegetarianism keeps the body and mind light, and helps us move towards self-realisation (Source: An Article Appeared in HT - March 2017).

Eating Habits Should Not Be Politicised: Hindutva brigade is that the cow is sacred and, therefore, cannot be slaughtered, geographical location and eating habits should not make a difference. To focus on eating habits is to open the floodgates to vigilantes who want to impose their own codes on people. The fatal fallout of that was seen in Dadri where a man was beaten to death in his own home on suspicion of storing beef in his fridge. The issue of eating habits should not even feature in election campaigns and the party should not favour anyone by saying that they would be allowed to eat whatever they want. What holds good for the northeastern states, irrespective of who they elect, should be so for other states as well (Source : HT 29.03.2017).

Opening Bank Accounts for Milk Producers: There are 1.7 lakh dairy cooperative societies (DCS) at village level, having around 1.6 crore milk producers, in the country. About 850 lakh litres of milk is procured per day from them. Value of milk procured from the DCS is to the tune of ₹120 crore per day. Opening of bank accounts of all those unbanked milk producers will protect farmers from cheating. The government should make policy and regulation conducive to help corporate participation in dairying. It will boost output directly and also spur innovation in the cooperative sector and further boost cooperativisation. Better is to form producer companies where there are no limits on the number of members and voting rights are based on the cooperative principle of one-man-one-vote. Producer companies are insulated from the institutional state and political interference in cooperatives through the office of the registrar of cooperatives.

Promoting Small holders' Dairying

Smallholders' dairying is a major activity, in India, which is holding the largest numbers, around 350 million farmers (70 million rural households), who keep dairy animals. Out of these 260 million milk farmers (52 million households) are small, marginal and landless. The herd sizes of these farmers vary from 1-2 to about 6. Amongst these farmers, around 65 million farmers (13 million rural households) are connected with the dairy cooperative institutions, in India. They have been socio-economically benefited due to an effective institutional framework, which provides them the best value of their produce. Besides they are supported with the technical input services like artificial insemination, veterinary services, feed supply, assured payment, bonus, credit facilities, etc. Other than the individual member benefits, the village community as a whole are being benefited by having roads, schools, hospitals, etc., built out of the surpluses generated by the milk cooperative institutions.

The balance 39 million small households of dairy animal, who are not connected with any effective institutional framework, are being faced with numerous constraints as mentioned above. These constraints have eroded the comparative advantages of the small holder's dairying in

India. The following policy measures will enable greater participation of the small holders.

- Improve infrastructures in rural areas.
- Introduce an effective disease control cum veterinary care, at an affordable price.
- Enhanced research and development, especially in the areas of low cost homegrown technologies.
- Create a favourable investment environment for smallholders.
- Awareness about product quality, hygiene and sanitation.
- Effective implementations of quality, hygiene and sanitary standards.
- Provide appropriate institutional framework for market access and best value realisation of the products.

Conserving Indigenous Breeds

According to the Food and Agriculture Organisation of the United Nations, India has a wide range of indigenous cow breeds. Some of them are teetering on extinction. While 69% of the indigenous breeds are with the poor, who cannot afford the fancy foreign breeds - Jerseys and Holsteins under the conditions such as reduced land for pastoral use, reduced need for bulls for agriculture, and lack of a policy specifically to conserve these breeds of cows. Local breeds are likely to sustain under uncertain whether droughts, heat and be able to feed on locally available feedstock better than other breeds and crossbreeds. We must ensure survival of such rich diversity. What should be the role of a Gaurakshak (cow protector)? Each Gaurakshak should take upon the task of lovingly nurturing and bringing up one or more cow breeds, ensuring they proliferate and their generic pool is conserved. That is the only type of Gaurakshak our cows need. Anything else is irrelevant and irresponsible. Their all efforts to curb cattle smuggling along the 4,096 km India-Bangladesh border has not made any change.

Differentiate Between Beef (Cow Meat) and Buffen (Buffalo Meat): The problem is the climate of fear Hindutva activists have created in the wake of beef bans in several states. Together with the decline in demand for Indian 'beef' (entirely buff) internationally, this has made a big impact on India's $4.8 billion 'beef' export business, which shrunk by over $700 million, or 15 percent, in the period from 2014-15 to 2015-16.

One of the factors finding play in the recent wave of cow vigilantism is the blurring of lines between beef and buff in the imagination of the majority. Moreover, buffalo meat is called "beef" in north India, where the complete ban on cow slaughter rules out any trade in cow beef. In Kerala, where cow slaughter is legal (with conditions), buffalo meat is popularly referred to as beef. 'Beef' is more appealing on the menu than 'buff'. Indeed, buffalo meat vendors and restaurants serving "beef" are feeling the heat. They know how vigilantes can target any meat trader, including buffalo meat suppliers, in the name of protecting cows.

Cow meat is supposed to be five percent of the total meat production in India, but such is the fear of cow-protection that you don't have to have anything to do with cow meat to fear being targeted. So widespread is the fear that restaurants nowadays request customers not to say the word beef aloud. Mumbai's largest abattoir in Deonar, which used to process over 1.5 lakh head of cattle a year, closed down, affecting the livelihood of butchers, meat processors, exporters and resellers across the country. Drought affected Maharashtra was the worst hit by the ban as farmers couldn't sell their cattle even in extreme distress. Prices fell by over 40 percent as soon as bulls and bullocks were brought under the ban, impacting 15 million livelihoods (Source: outlook-19 Sept. 2016).

Legalise Cattle Trade to Curb Smuggling

The only way to tide over the problem of cattle smuggling is by "legalizing export of over-aged and non-milching animals" and "promoting beef industry in India" by allowing their slaughter. It seems unpalatable at a time when some states have banned cow slaughter. It is time to take a practical approach, albeit with respect for the cultural sensitivity of the country.

Other suggestions are opening cattle haats in bordering areas for limited sale of cattle to Bangladesh; legalizing cattle export for agricultural purposes and helping Bangladesh develop animal husbandry and total ban on slaughter of cattle in India by enactment of law by the government of India.

However, if total ban on slaughter of cattle is imposed, we may face problems such as we do not have enough infrastructure, feed, fodder and grassland to support overaged and non-milching cattle heads therefore imposing complete ban on slaughter of cow may attract strong objections from many corners.

Disrupting Cattle Markets is Economically Unviable

Last year in May, 2017 legislation represented an attempt to impose cultural beliefs of a section of society on the dairy industry. The economic burden of this attempt was born by farmers. Even Prome Minister talked about steps to lighten the regulatory environment for doing business in the non-agricultural sector which was subjected to additional burdens. For India's farmers, this came at a time when they had to confront a new danger, climate change, quite apart from the usual problems of low productivity, land fragmentation, lack of access to credit and markets.

Cattle are an asset for a farmer, with a visible production life cycle. Once an asset is no longer productive for an owner, it is only natural that he finds another buyer who can subsequently put it to use. Once this practice is disrupted, farmers and cattle traders bore the economic burden. In additions, non-state actors in the form of vigilante groups have also been allowed to harm trading networks. Ironically, states with strict slaughter bans typically have a higher proportion of strays as farmers turn them loose, unable to bear the burden of feeding them. The government should change its approach to the farm sector; try to create a national online market to give farmers more options. Why to shakle them by disrupting cattle markets. Agriculture needs ease of doing business as much as other economic sectors.

Annexure-1

Acronyms

AIDIS	:	All India Debt and Investment Survey
APEDA	:	Agricultural Products Export Development Authority
AWB	:	Animal Welfare Board
BCE	:	Before Christ Era
BCM-7	:	Beta-Casomorphin-7
BCS	:	Body Condition Score
BISA	:	Borlaug Institute for South Asia
CBI	:	Central Bureau of Investigation
CCBF	:	Central Cattle Breeding Farms
DCS	:	Dairy Cooperative Societies
DTE	:	Down to Earth
EIA	:	Environment Impact Assesement
ETT	:	Embroy Transfer Technology
FPR	:	Field Performance Recording
FSDA	:	Food Safety and Drug Administration
GDP	:	Gross Domestic Product
GRD	:	Gau Rakshak Dal
HACCP	:	Hazard Analysis and Critical Control Points

IGRFI	:	Indian Grassland and Fodder Research Institute
MCD	:	Municipal Corporation of Delhi
MIS	:	Management Information System
MOE	:	Ministry of Environment
MSP	:	Minimum Support Price
MT	:	Million Tone
MTC	:	Micro Training Centre
NAAS	:	National Academy of Agricultural Sciences
NDDB	:	National Dairy Development Board
NDMC	:	New Delhi Municipal Corporation
NOC	:	No Objection Certificate
NPCBB	:	National Project for Cattle and Buffalo Breeding
NSSO	:	National Sample Survey Office
OPU-IVF	:	Ovum Pickup-Invitro Fertilisation
SHGs	:	Self Help Groups
SIA	:	State Implementing Agency
SMS	:	Straw Management System
SVAROP	:	Scientific Validation and Research on Panchgavya
TDN	:	Total Digestible Nutrients